A Brief Introduction to the Search for Extra-Terrestrial Life

A Brief Introduction to the Search for Extra-Terrestrial Life

By
Cesare Barbieri
Professor Emeritus of Astronomy, University of Padova

CRC Press
Taylor & Francis Group
Boca Raton London New York

CRC Press is an imprint of the
Taylor & Francis Group, an **informa** business

CRC Press
Taylor & Francis Group
6000 Broken Sound Parkway NW, Suite 300
Boca Raton, FL 33487-2742

© 2019 by Taylor & Francis Group, LLC
CRC Press is an imprint of Taylor & Francis Group, an Informa business

No claim to original U.S. Government works

Printed on acid-free paper

International Standard Book Number-13: 978-0-367-19194-8 (Hardback)

Library of Congress Cataloging-in-Publication Data

LoC Data here (or delete if not supplied)

Visit the Taylor & Francis Web site at
http://www.taylorandfrancis.com

and the CRC Press Web site at
http://www.crcpress.com

To Lavinia, Stefano, Leonardo, Tommaso

Contents

Acknowledgments

This review originates from the lectures given by the author to the School of Higher Education "Giacomo Leopardi" of the University of Macerata and the Shanghai Tech University—University of Padova Summer, at the invitation of their Directors, Professors Luigi Alici and Ernesto Carafoli, respectively.

Dr. Paolo Ocher made several suggestions and provided images and spectra from the Asiago Astrophysical Observatory.

Dr. Ivano Bertini made useful suggestions to improve the text. I enjoyed many conversations on prebiotic processes with Dr. Mario Casarotto.

The reading of the first draft of the paper by Andrea J. Tommaso and Stefano Barbieri—helped to improve the legibility of the content.

List of Abbreviations

ASI	Italian Space Agency
AU	Astronomical Unit of distance
C	Degree Centigrade
C-G	Churyumov – Gerasimenko
CNSA	Chinese National Space Agency
DNA	deoxyribonucleic acid
ELT	European Large Telescope
EPFL	École Polytechnique Féderale de Lausanne
ESA	European Space Agency
ESO	European Southern Observatory
eV	electron Volt (keV, MeV)
FAST	Five hundred m Aperture Spherical Telescope
HARPS	High Accuracy Radial velocity Planet Searcher
HST	Hubble Space Telescope
HZ	Habitable zone
IAU	International Astronomical Union
INAF	Italian National Institute for Astrophysics
IRAM	Institut de RadioAstronomie Millimetrique
ISO	Infrared Space Observatory
ISS	International Space Station
JAXA	Japan Aerospace Exploration Agency
JFC	Jupiter Family of Comets
JWST	James Webb Space Telescope
K	Degree Kelvin
KARI	Korean Aerospace Institute

L2	second Lagrangian point
LHB	Late Heavy Bombardment
LRO	Lunar Reconnaissance Orbiter
LUCA	last common universal ancestor
l-y	light-year, unit of distance (Ml-y, Gl-y millions, billions of light-years respectively)
MK	Morgan and Keenan
MRO	Mars Reconnaissance Orbiter
NAI	NASA Astrobiology Institute
NASA	National Aeronautics and Space Administration
OAPd	Astronomical Observatory of Padova
OPR	ortho to para ratio
pc	parsec, unit of distance (kpc, Mpc, Gpc, thousand, million, billion pc respectively)
RNA	ribonucleic acid
SN	Supernovae
TESS	Transiting Exoplanet Survey Satellite
TNG	Telescopio Nazionale Galileo
TNOs	trans-Neptunian objects
Unipd	University of Padova
VSMOW	Vienna Standard Mean Ocean Water
y	years (ky, My, Gy; thousands, millions or billions of years respectively)

CHAPTER 1

Introduction

W E LIVE IN A period of extraordinary discoveries in our
Solar System, in the planetary systems of other stars (exoplanets) and in the complex molecules in interstellar and intergalactic clouds. These advancements promote new investigations of
the meaning of "life."

Our knowledge about life developed over the centuries thanks
to the many philosophers, physicists, chemists and biologists,
who examined such complex matters according to their different
points of view. Out of this long history, I wish to quote here only
one date, the year 1953. In that year, Miller and Urey carried out
their famous experiment about the primordial universal soup,
whose foundations had already been expounded by the Russian
chemist Alexandre Oparin in 1924. From a mixture of five gases,
methane, ammonia, carbon dioxide, hydrogen and water vapor,
and an electric discharge as the source of energy, complex molecules were produced, including amino acids. In the same year,
James D. Watson, Francis Crick and collaborators discovered
the double helix of DNA (deoxyribonucleic acid), a nucleic acid
containing the genetic information needed for the biosynthesis
of RNA (ribonucleic acid) and proteins. DNA, RNA, proteins

and carbohydrates are the main macromolecules essential for all known living beings.

As a source of inspiration for their researches, both Watson and Crick acknowledged the book by Erwin Schrödinger, *What Is Life?* (1944). In that book, Schrödinger recognized that the question of life requires a multidisciplinary approach. The fundamental importance of quantum mechanics was also pointed out, and far-reaching questions were formulated such as the apparent violation of the second principle of thermodynamics. Life seems to be characterized by a sort of negative entropy, a "negentropy" in his word. See the review of Schrödinger's book by Ball (2018). Today, we might add another question: How crucial is information for life? In an often-quoted sentence, the famous physicist Archibald Wheeler said, "It from bit," be the bit a classic or a quantum one, namely from information to matter. Is information a free-floating entity such as electromagnetic or gravitational fields? Does entropy have an entirely different meaning as suspected by Schrödinger? Is life analogic or digital, or both at the same time, as in peptides, digital in the sequence of their amino acids and analogic in the complexity of their macrostructures? These and other questions of a profoundly philosophical nature exceed the purpose of the present astronomically oriented review. See, for instance, the many papers devoted to the role of information on life contained in the book by Walker et al. (2017).

For the past 50 years, the space age has allowed us to study life not only on Earth but also on other planets of the Solar System and their moons, in a variety of extra-terrestrial environments and on exoplanets. A major leap was made more or less at the same time as the primordial soup experiment and the identification of DNA, when Earth was considered and studied as any other planet; thus, comparative planetology flourished. Today, the same step is being made by comparing our Solar System to the planetary systems of other stars.

To study the broad subject of life on Earth and elsewhere, a new discipline has emerged, astrobiology, which involves not only scholars of astronomy, physics, chemistry, biology and engineering, but also sociologists, philosophers and theologians, and arouses the interest of the whole society. The NASA Astrobiology Institute (NAI, https://nai.nasa.gov/) was established in 1998. The NAI is a virtual, distributed organization of teams that integrate astrobiology research and training programs in concert with the national and international science communities. In Europe, a European Astrobiology Institute is proposed (http://europeanastrobiology.eu/), to be finalized in 2019. See the book by Capova et al. (2018) on this topic.

The present review will discuss several topics and questions raised by the fascinating subject of terrestrial and extra-terrestrial life, with the aim of providing at least some information and answers from the astronomical point of view of the author. The review is organized across several chapters.

Chapters 2 and 3 illustrate how the Universe's chemical composition evolved, from the initial conditions after the Big Bang, to the present abundances of the different elements thanks to nuclear reactions inside the earliest stars. The evolution of large aggregations of matter, namely galaxies, their clusters and the intergalactic clouds of dust and gases, led to the current structure of the Milky Way and the Solar System within.

Chapters 4 and 5 treat questions such as how did life originate? When and where on Earth? Did cosmic bodies such as comets and asteroids have any influence on the appearance of the first building blocks of life? What are the main characteristics of life that will allow us to recognize it in extra-terrestrial environments? The basic characteristics of living organisms and the paramount importance for life of the three phases of water (solid, liquid, gaseous) will be examined. The presence of those phases in many bodies of the Solar System, in interstellar matter and most likely on exoplanets will be discussed.

Chapter 6 discusses the perspectives and difficulties of establishing human bases outside Earth. Instinctively, we tend to associate the adjective "extra-terrestrial" with the life of other living and perhaps intelligent beings somewhere in the universe. Actually, the concept of extra-terrestrial life has another implication, namely the possibility to export humanity to other worlds for permanent outposts, in particular to the Moon and Mars, in the not too distant future.

Chapter 7 reviews the current knowledge on several extra-terrestrial environments that host the building blocks of life or perhaps are capable of supporting life, based on data provided by ground and space telescopes such as the Atacama Large Millimeter/Submillimeter Array (ALMA), the Hubble Space Telescope (HST), Gaia, Kepler and the Transiting Exoplanet Survey Satellite (TESS), and in situ missions such as Rosetta (discussed in Chapter 8). In addition to our instruments, nature itself provides precious information about cosmic matter, thanks to the continuous infall of meteorites, some of them coming from the Moon or Mars or the asteroid Vesta.

Chapter 8 is devoted to the European cometary mission Rosetta. For the first time in space exploration, a spacecraft spent almost two years inside the coma of a comet, from before to after the perihelion. The high number of instruments aboard (imaging devices, spectrographs, thermal mappers, magnetometer, mass spectrometers, etc.) provided a unique amount of information about the morphological and geological surface, the magnetic field and the chemistry of the atmosphere of a comet. Various points in this review quote some of the results obtained by Rosetta, while Chapter 8 will detail only some of the results directly relevant to life.

Chapter 9 illustrates the current evidence about exoplanets, their number and location and possible habitability according to the characteristics of the host star. Furthermore, the main discovery techniques by ground and space telescopes will be exposed.

Chapter 10 examines the possibility of finding intelligent species around us by radio or optical means. The Fermi paradox and the Drake equation will be discussed in such context. The symmetric event, of our Earth being discovered by aliens, will be briefly examined.

The concluding chapter, Chapter 11, will draw some conclusions, with the proviso that the field is in tumultuous expansion, with new discoveries and new theories frequently published. Hopefully, a number of notions in the current review will survive the passage of time.

To conclude this introduction, we underline that the developments in knowledge about our Earth, the Moon, Mars, the moons of Jupiter and Saturn, the many objects populating the outer Solar System and the discovery of planets orbiting other stars, have prompted the publication of a large number of excellent review papers and books. In a very personal and incomplete selection, the following books can be quoted: Goldsmith and Owen (2002), Di Mauro and Saladino (2016, in Italian) and Cockell (2015, 2017). An excellent textbook is *Life in the Universe* by Bennett and Shostak (2007).

Regarding reviews, Cottin et al. (2017a) summarize the results of a European Space Agency (ESA) topical team of interdisciplinary scientists focused on the utilization of near-Earth space for astrobiology. This paper presents an overview of past and current research in astrobiology conducted in Earth's orbit and beyond, with a special focus on ESA missions such as Biopan, STONE (on Russian FOTON capsules) and EXPOSE facilities (outside the International Space Station). In an accompanying paper, Martins et al. (2017) review how Earth is used by astrobiologists to investigate life in extreme environments, its metabolisms, adaptation strategies and biosignatures; extinct life from the oldest rocks on our planet and its biosignatures; and changes in minerals, biosignatures and microorganisms under peculiar space conditions.

Several reliable websites are quoted throughout the book because information is quickly updated thanks to the policy adopted by many agencies to distribute their results as soon as they are verified. Some examples follow and other quotations can be found in the relevant chapters.

The International Astronomical Union (IAU) has established a working group on Education and Training in Astrobiology that has prepared a platform of online courses in astrobiology designed for upper-level graduate students and postdocs, as well as the interested and curious public (see http://astrobiovideo.com/en/). A very useful site for information on the Solar System, including astrobiological matters, is the Solar System Exploration Resource Virtual Institute (SSERVI, https://sservi.nasa.gov/). See https://www3.nd.edu/~cneal/Lunar-L/ for a repository of lunar-related documents and links. For the earlier Soviet Union missions to the Moon, see https://www3.nd.edu/~cneal/Lunar-L/4_Lunas_Lunokhods_Macau_sm.pdf. The survey of the scientific literature reported in the references was carried out to the end of 2018.

CHAPTER 2

Cosmological Process Leading to Life on Earth

I T ALL STARTED WITH the so-called "Big Bang," namely with the beginning of the expansion of the Universe, approximately 13.8 billion years ago.[1] The following notations will be used: Gy and My to indicate billions and millions of years, respectively; cosmological distances will be indicated mostly in light-years, l-y, so Gl-y, Ml-y. Notice that many papers use the notation Ga for gigayears, not Gy as here.

As first proposed by Alpher and Gamow around 1948, when the expanding Universe cooled down and reached temperatures equivalent to the binding energy per nucleon, namely billions of degrees Kelvin (in physical units energies of about 10 million electron volts, MeV), nucleosynthesis occurred. The rapid cooling allowed the formation of the first elements through fusion reactions using protons p (nuclei of hydrogen H, indicated by H^+) and neutrons n present in the environment. The main processes were the p–p reaction producing deuterium (D), and the D–D reaction producing helium

(H^3 and H^4). In addition, minor channels such as $D + He^3$, $p + He^3$ and $n + He^3$ were also present. The light elements lithium (Li^6 and Li^7) and beryllium (Be^7) were produced in minute quantities by subsequent processes involving H, He^3 and He^4. The primordial nucleosynthesis essentially stopped at Li^7, because there are no stable elements with atomic mass $A = 5$ or $A = 8$ (see, for instance, the book by De Angelis and Pimenta, 2018).

In essence, around 75% of the mass of primordial ordinary matter (in physical terms, baryonic matter) was H, 25% was He^4, with tiny fractions of D, He^3, Li^7 and Be^7. A first generation of stars produced heavier elements through nuclear fusion processes in their interiors, where temperatures ranged from 108 K to 106 K, with heavier stars being hotter. After no more than 1 Gy, the formation of galaxies, their clusters, stars and perhaps planets, was already underway, although the star-formation rate had its peak at around 3 Gy after the beginning (at cosmological redshifts of z around 2),[2] as was recently confirmed by the Fermi–LAT collaboration using gamma-ray data (Ajello et al., 2018).

Such an evolution of structures was accompanied by an equally fast chemical evolution. The explosions of first-generation stars as supernovae (SN) enriched the interstellar medium of heavy elements, so that successive generations of stars had those elements and more complex nuclear reactions at their disposal. To be more precise, there are two basic types of SN.

Type I SN occur in binary systems with at least one star of approximately one solar mass, a surface temperature of around 104 K and a radius comparable to that of Earth, namely a so-called white dwarf, the prototype being the faint companion of Sirius. A distinctive feature of Type I SN is the absence of hydrogen lines in their spectra. Type I SN belonging to the subclass "Ia" are the best standard candles with which to measure the expansion of the Universe and provide firm evidence of the existence of the dark energy accelerating the expansion.

On the other hand, Type II SN show hydrogen lines in their spectra and result from the rapid collapse of a star of at least eight solar masses. See, for instance, the very informative article about the elements dispersed by the Type II SN in Cassiopeia A given by the Chandra X-ray Observatory website http://chandra.si.edu/photo/2017/casa_life/. The Cassiopeia A remnant contains large quantities of hydrogen, carbon, nitrogen, oxygen and phosphorus, namely the elements required to make DNA and thus the building blocks of life.

The details of the chemical evolution of the Universe are very complex and depend not only on the cosmological epoch but also on the environment where the star or the nebula was located. In a drastic approximation, we can say though that after less than 1 Gy all main chemical elements were present, both in the stars and in the interstellar nebulae. See, for instance, the paper by Hashimoto et al. (2018), which used data from the ground telescopes Very Large Telescope and Atacama Large Millimeter/Submillimeter Array (VLT and ALMA, respectively; both telescopes are described in https://www.eso.org/) and the Hubble Space Telescope (HST; http://hubblesite.org/), showing that the stars of the very distant galaxy MACS1149-JD1 were formed approximately 250 My after the Big Bang. The galaxy is a member of a cluster with a redshift of $z = 9.6$ (namely a distance of about 13 Gl-y) and is contoured by a huge cloud of ionized oxygen. Therefore, this galaxy is the most distant known source of oxygen.

As already pointed out, the percentages of the elements varied in a complicated manner with subsequent generations of stars and the ambient in which the stars were born. According to a paper by Noguchi (2018), even the history of star formation in the neighborhood of the Solar System may have been more complicated than believed until recently. The solar neighborhood shows indeed the existence of two distinct groups of stars, one with low iron (Fe) content and the other with high Fe content (low and high with respect to the solar composition). The author concludes that the low Fe stars formed first in the very initial phases, while the high Fe stars formed during a second phase. The peaks in these two star-formation episodes were separated by approximately 5 Gy.

Regarding chemical abundances, our Sun is the main benchmark. Table 2.1 shows the photospheric solar composition for selected elements, in a $\log_{10}$ scale, conventionally normalized to 12 for H. The composition is dominated by H, He, C, N, O and so on in decreasing order until the strong peaks of Fe and nickel (Ni) are reached. See Asplund et al. (2009) for a detailed discussion of Table 2.1. To be more precise, our knowledge of the overall abundance distribution of the elements in the Solar System originates from two quite different sources: the solar photospheric spectrum

TABLE 2.1   Solar Photospheric Abundance (by Number) of Selected Chemical Elements

Symbol	Atomic Number	Atomic Weight	Log Abundance by Number
H	1	1.008	12.00
He	2	4.003	10.9
Li	3	6.941	1.1
Be	4	9.012	1.4
B	5	10.811	2.7
C	6	12.011	8.4
N	7	14.007	7.8
O	8	15.999	8.7
F	9	18.998	4.6
Ne	10	20.179	7.9
Na	11	22.9898	6.2
Mg	12	24.305	7.6
Al	13	26.9815	6.5
Si	14	28.086	7.5
P	15	30.974	5.4
S	16	32.06	7.1
Ar	18	39.948	6.4
Ca	20	40.08	6.3
Mn	25	54.9380	5.4
Fe	26	55.847	7.5
Ni	28	58.71	6.2
Mo	42	114.82	1.9
Xe	54	131.30	2.2

and the pristine meteorites, in particular the so-called C1 chondrites.[3] Meteoritic abundances, including their isotopic content, can be measured to exquisite accuracy in the laboratory, with the caveat that elements such as H, He, C, N, O and neon (Ne) are all volatile and hence depleted in meteorites. Table 2.2 provides the same information on a linear scale normalized to 100 for H, grouping the elements together. As usual in astronomical jargon, the category "metals" comprises all other elements.

As already mentioned, hydrogen came entirely from the Big Bang, as did deuterium and almost all helium. Carbon, nitrogen, oxygen and the other elements until the iron–cobalt–nickel group[4] were made by nuclear fusion processes in the interiors or atmospheres of different types of stars. Until recently, much heavier elements such as gold and platinum represented a puzzle, because their formation requires the presence of many neutrons in the ambient. In 2017, the detection of short-duration gamma-ray bursts accompanied by a gravitational wave event, with a transient optical-infrared counterpart, provided an additional formation mechanism. The merging of two neutron stars produced a kilonova (or meganova), whose spectra showed the presence of such heavy elements (see Pian et al., 2017, Covino et al., 2017).

Notice that H, C, N and O are the main chemical ingredients of bacteria and our own body, as discussed in more depth in Chapter 4. Therefore, excluding inert gases such as He, the composition of living beings is much more similar to that of stars and the

TABLE 2.2   Relative Abundance by Mass and by Number, Normalized to 100 for Hydrogen, Grouping Together the Main Elements, where "Metals" Means all the Rest

Element Group	Mass	Number
H	100.00	100.000
He	34.00	8.500
C, N, O, Ne	1.75	0.116
Metals	0.50	0.014

interstellar medium than that of the Earth. Indeed, the Earth's crust is rich in silicon (Si, 10 times more abundant than carbon, and the preferred element by science-fiction writers for alternative forms of life) and heavier elements, while the terrestrial atmosphere is deprived of hydrogen and carbon (99% of which is present as carbon dioxide, CO_2).

As Carl Sagan (1934–1996), a founder of astrobiology, said:

The nitrogen in our DNA, the calcium in our teeth, the iron in our blood, the carbon in our apple pies were made in the interiors of collapsing stars. We are made of star stuff.

The long-held theory of "panspermia," discussed two centuries ago by illustrious scientists such as H. von Helmholtz, Lord Kelvin, S. Arrhenius and others, seems to find confirmation in modern data. Hence, we can postulate that life is diffuse in the whole Universe and maintain the reasonable hope that we will be capable of recognizing its signature in extra-terrestrial environments and distant alien planets, should life actually be there. These points will be discussed in Chapters 9 and 10.

Not all is clear though. First, there is the problem of deuterium (D), the only stable isotope of H and of astrophysical interest. The Big Bang produced a tiny amount of D, as confirmed for instance by the very low D/H ratio in the solar nebula (approximately 10^{-5}; Geiss and Reeves, 1972). Nuclear processes in the interior of stars do not produce new D, because it appears only as an intermediate product. Deuterium can be synthesized in the atmospheres of active stars (for instance, production of D has been detected in solar flares), but such production of fresh D is so minute as not to change the main point of its small abundance with respect to H. However, observations show that the values of D/H in the interstellar medium, planetary atmospheres, terrestrial water, comets, etc., span a factor of more than 50, raising in some cases the value of D/H well above the standard terrestrial value of 1.5 $\times 10^{-4}$ (see also later in this chapter and as further discussed in Chapter 8 when speaking of comet 67P). Therefore, enrichment or fractionation processes are required to explain the wide range.

Theoretical studies of interstellar chemistry show that in astrophysical environments, at temperatures lower than approximately 50 K, water becomes enriched in deuterium relative to molecular hydrogen. See, for instance, Owen et al. (1986) for a seminal paper on deuterium in the outer Solar System, where two distinct reservoirs of primordial D were identified, one contained in gaseous hydrogen and the other contained in low temperature volatiles or trapped in cold, solid material isolated from hydrogen. Both reservoirs were present before the formation of the Solar System. Hallis (2017) gives a review of the D/H ratios in the inner Solar System, which appears to be relatively homogeneous in terms of D/H. It is worth recalling the D/H value in the so-called Vienna Standard Mean Ocean Water (VSMOW) model: $D/HSMOW = 1.5576 \times 10^{-4}$. There are deuterium-enriched reservoirs such as the interstellar water ices and D-poor reservoirs such as the protosolar disc. According to Hallis, the original water D/H ratios of Earth and Mars and asteroids such as Vesta were slightly D-poorer than VSMOW. On the other hand, the Rosina instrument of the cometary mission Rosetta found an appreciable enrichment in the gaseous coma of comet 67P, as discussed in Chapter 8.

Another puzzle not fully solved is the under-abundance of the light elements Li, Be and boron (B) with respect to their neighbors on the periodic table, H and He on the one side, and C, N and O on the other. Although Li, Be and B are formed by nuclear fusion in the deep interior of stars, they are also destroyed by other reactions close to the surface. However, these elements are needed for RNA and DNA. Boron is particularly important in the present context, because borate anions (BO_4^-) may be necessary for the origin and maintenance of life. The reason is that borates stabilize ribose, the simple sugar that forms the backbone of RNA with phosphate. Without borates, ribose quickly decomposes in water. Although other methods are proposed for producing RNA, borates may have been a necessary bridge from abiotically produced organic molecules to RNA-based proto-life on Earth. A recent finding (Gasda et al., 2017) claims the detection of B traces by drilling a Martian deposit on the Gale crater. This finding further opens the possibility that life could have once arisen on the red planet. As an aside consideration, this and other discoveries underline the great importance of studying matter, as a channel of information parallel to the electromagnetic channel. The two channels together will complement and enlarge our knowledge about the universe. At the time of writing, the Japanese mission Hayabusa 2 to asteroid Ryugu (http://www.hayabusa2.jaxa.jp/en/) and NASA mission Osiris-Rex to asteroid Bennu (https://www.nasa.gov/osiris-rex) are underway, both with the aim to bring back to Earth samples of material collected on the surface of those two pristine bodies. Hayabusa 2 and Osiris-Rex are the first such missions since the Apollo and Lunokhod enterprises of 50 years ago. More sample-return missions from a variety of Solar System bodies are badly needed!

Returning to Li, Be and B, additional formation mechanisms must be found. The bombardment of heavy atoms by high-energy particles called "cosmic rays," capable of splitting them into lighter ones in a process called "spallation," appears to be an important source of Be and B (for Li, a partial nuclear source is viable).

Another element whose low cosmic abundance causes some problems for life is molybdenum, which seems to be essential for the uptake of nitrogen from both nitrogen gas and nitrate. See the discussion by Williams and Fraústo da Silva (2002). A paper discussing such problems in connection with the last common universal ancestor (LUCA, see Chapter 4) is by Schoepp-Cothenet et al. (2012).

NOTES

1. Dark matter and dark energy are not discussed in this review because they do not seem to have a direct link to emergency and properties of life.
2. The expansion of the universe shifts the wavelength λ of all atomic and molecular spectral lines and bands toward the red by an amount $\Delta\lambda$ increasing with the distance of the emitter from the observer. The quantity $z = \Delta\lambda/\lambda$ is the cosmological redshift.

3. Meteorites are cosmic objects of cometary or asteroidal origin. Those reaching the ground belong to several petrographic varieties, ranging from carbonaceous to metallic. Carbonaceous types (C-type, chondrites) are the most commonly found. The C1 variety is the least modified by various physical and chemical processes over the past 4.6 Gy, thus containing pristine information.
4. The nuclear reactions after the Fe–Co–Ni elements become endothermic, thus requiring modifications to the stellar structure.

CHAPTER 3

Events in the Milky Way and Solar System

L ET US PUT THE previous considerations in the context of the Solar System as a member of the Milky Way, our own galaxy. Terms such as "comets," "asteroids," "meteorites" and "interplanetary dust," will be used to indicate the "minor" (regarding size, not importance) constituents of our Solar System and most likely of the planetary systems of other stars. The term "planetesimal" indicates the first aggregates of matter in the original solar nebula whose size of several kilometers allowed mutual gravity to play the primary role in their successive coalescence into larger bodies. Distances inside the Solar System and the realm of the nearer stars will be given in Astronomical Units (AUs, the average distance Earth–Sun, 1 AU $\approx 1.5 \times 10^8$ km), in parsec (pc, 1 pc $\approx$ 206,265 AUs) or in light-years (l-y, 1 pc $\approx$ 3.26 l-y).

The Solar System is a component of a spiral galaxy (the Milky Way), residing in its symmetry plane at a distance of approximately 8.1 kpc ($\approx$26,500 l-y) from its center, which is occupied by a black hole of about 4 million solar masses. The overall rotation takes the Sun to span a narrow torus around this center, with a revolution period of about 225 My (one galactic year). Therefore, using the value of 4.6 Gy for the solidification of the Earth's crust, derived from the analysis of the lunar rocks brought back by the Apollo and Lunokhod missions of 50 years ago, life on Earth is only 20 galactic years old. In addition to the revolution, the galactic field of forces produces a pendulum motion of the Solar System above and below the plane of the Milky Way, reaching a maximum distance of a few tens of parsec every 30 My or so. Moreover, the Solar System has a proper motion with respect to nearby stars with a velocity of about 20 km/s, approximately in the direction of the bright star Vega. As a consequence of such crossing of the galactic plane and proper motion, the Solar System experiences a varying environment of stars, gases and dust, with the possibility to enrich its molecular content every time it encounters an interstellar cloud. To be sure, the story of those 20 galactic years is largely to be written. Surely, it was not a quiet ride around a distant center. Everything was moving and changing, star clusters formed and dissolved, and so did the great interstellar nebulae. Stars exploded and modified the chemistry, the black holes of stellar mass coalesced and the Solar System never really went back to the same galactic position. Even the galaxies around the Milky Way such as M 31 in Andromeda, the Magellanic Clouds and so on, were not how nor where we see them today. The European astrometric satellite Gaia (http://sci.esa.int/gaia/), after 22 months of operation, already provides data that lead to an unexpected scenario. Instead of forming alone, our galaxy merged with another large galaxy early in its life, around 10 Gy ago (Helmi, 2018). According to this paper, all around the sky there are large numbers of stars left from an object (nicknamed Gaia-Enceladus by the authors) that at infall had a mass about one-fourth of the Milky Way. Such a result was made possible by the exquisite precision of Gaia to measure the positions, proper motions and colors of a large number of stars, which could then be placed accurately in a multidimensional space of galactic coordinates, velocities, chemical compositions and ages. Gaia, the successor of the Hipparcos satellite that operated 20 years ago, is located in the L2 Lagrangian point of the Sun–Earth system. The Lagrangian points are so named in honor of the Italian-French mathematician Giuseppe M. La Grangia (in French, Joseph M. Lagrange) who discovered their existence by studying the motion of a point of negligible mass (e.g., a comet or artificial satellite) in the gravitation field of two more massive bodies, e.g., Sun–Earth or Earth–Moon or Sun–Jupiter. If a point of small mass is put in one of those five points, its position will remain stable with respect to the rotating line joining the two larger masses. Actually, the stability is not perfect due to the influence of perturbations, e.g., by the solar radiation pressure; nevertheless, the satellite will describe for a long time a complex but not too large orbit around the point.

For clarity, Figure 3.1 shows the position of the Lagrangian points of the Sun–Earth system. Point L1, located at a distance of approximately 1.5×10^6 km from Earth, as L2, is the ideal place for continuous monitoring of the solar activity (the satellite SoHO is parked there). Points L4 and L5 are at the vertexes of equilateral triangles and may harbor rarefied and temporary clouds of dust. Point L3 is unobservable.

Point L2 is particularly important in the context of the present review because it is used as a parking place for several artificial satellites, which will be mentioned again in Chapters 6, 7, 9 and 10.

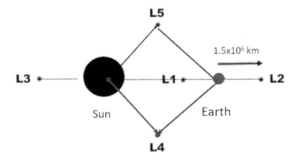

FIGURE 3.1 The Lagrangian points in the Sun–Earth system.

Understanding the complex series of events interesting our Milky Way and the nearer galaxies, our Solar System and the Earth within it, will greatly benefit from the data of the European astrometric satellite. The importance of Gaia for exoplanet studies will be discussed in Chapter 9.

3.1 THE STRUCTURE OF THE SOLAR SYSTEM

Coming now to our Solar System, Emanuel Swedenborg in 1734 was perhaps the first to formulate sound ideas on how the Solar System grew out of a rotating, contracting and flattened nebula. Twenty years later, the great German philosopher Immanuel Kant reconsidered Swedenborg's hypothesis. Finally, in 1796, the French mathematician Pierre-Simon de Laplace added more physics and mathematics. Their theories constitute the so-called Kant–Laplace model of the eighteenth century, which maintains elements of validity even today. The contraction of the original nebula produced a differentiation of materials and temperature around the proto-Sun. At the end of a very complex evolution, the temperatures from Mercury to Mars allowed the formation of rocky planets with a high metallic content; those from Jupiter to Uranus produced gaseous planets with a high content of water ice, while further out the condensed bodies contained, in addition to water, high quantities of CO_2, CH_4 and N_2 ices.

According to recent theories (the so-called *Nice* model, see e.g., Tsiganis et al., 2005, Levison et al., 2008), the initial configuration of the Solar System differed significantly from today. The distance and even the order of the planets had most dramatic and rapid changes. Jupiter and Saturn moved outside and increased their distances and eccentricities, and Uranus and Neptune had their positions interchanged. Regions full of debris were left too, such as the Asteroid Main Belt in between Mars and Jupiter, the Kuiper Belt beyond Neptune and the Oort cometary cloud at the outskirts of the system.

What happened in the solar nebula and primordial disk prior to the giant planet orbit instability foreseen by the *Nice* model has been discussed by several authors. For instance, Davidsson et al. (2016), in the frame of the Rosetta mission to comet 67P, propose that most comet nuclei are a primordial rubble-pile, not collisional rubble piles formed in the aftermath of catastrophic disruptions of larger parent bodies. They argue that trans-Neptunian objects (TNOs) formed at sizes below ~400 km and that ~350 of these grew slowly in a low-mass primordial disk to the size of Triton, Pluto and Eris. Therefore, the primordial disk was dynamically cold, thus preventing medium-sized TNOs from breaking into collisional rubble piles and allowing the survival of primordial rubble-pile comets.

Knowledge of the structure of the Solar System beyond 50 or 60 AUs remains limited. The reason is that the observer on Earth sees the apparent luminosity of the distant bodies decreasing with the fourth power of the distance from the Sun. Moreover, the body reflects only a percentage of the incoming solar flux, a percentage indicated by the term "albedo." The albedo varies within a fairly wide range, e.g., 60% for a cloud-covered Earth to 7% for the Moon to 5% for a cometary nucleus. Given their distance from the Sun, the TNOs are very cool bodies, and observing in the infrared gives a great advantage over the conventional optical band. Therefore, only the largest telescopes equipped with suitable detectors collect a sufficient photon flux. Future observations with 30–40 m-class ground telescopes (e.g., the European Extremely Large Telescope [ELT]) and NASA's James Webb Space Telescope (JWST) will surely add significant new knowledge.

At the time of writing (January 2019), the most distant cosmic object known (discovered with the Japanese 8m Subaru telescope in Hawaii, nicknamed Farout, with a provisional designation 2018VG18 from the International Astronomical Union [IAU]) is located 120 AU from the Sun, and its orbit may take more than 1000 years.

With great admiration for the ingenuity and courage of the teams responsible, we are reminded that those remote regions of the Solar Systems are crossed by artificial bodies, the Pioneers and Voyagers launched decades ago (more about them in Chapter 10) and by New Horizons. On January 1, 2019, New Horizons flew by the Kuiper Belt object 2014 MU69 and sent back the first resolved image of such distant object (http://pluto.jhuapl.edu/). This TNO, at a distance of 43 AUs from the Sun, has been renamed Ultima Thule.

From time to time, the long-suspected presence of a ninth large mass planet (also referred to as Planet X) is announced, but at the moment, there is no definitive evidence for its existence.

Possibly, the outer structure was influenced by a passing star (see Pfalzner et al., 2018). The star, having a mass comparable to that of the Sun, approached the Sun at about three times the distance of Sun–Neptune, stealing most of the outer material from the protoplanetary disk and throwing what was left into inclined and eccentric orbits. These results seem to explain several features of the outer Solar System, including the mass ratio between Neptune and Uranus, and the existence of two distinct populations of Kuiper Belt objects.

Regarding the initial inner Solar System, it is estimated that it was populated by around 50–100 protoplanets ranging in size from the Moon to Mars that could experience giant collisions. Bodies that formed within the orbit of Mars contained essentially no water, as the conditions were too hot for volatile material to condense. Water was delivered from outside this region via a sequence of collisions. Progressively, the impactor population was dominated by smaller and smaller fragments, down to millimeter sizes. Among these later impactors, the relative importance of asteroids and comets as carriers of water ice and organic compounds is still debated (Rickman et al., 2017, Barbieri and Bertini, 2017).

3.2 THE IMPORTANCE OF THE MOON

The Moon is of paramount importance in contemporary everyday life for several reasons: from its influence on climate (e.g., Camuffo, 2000) to the many processes that affect the rhythms of animals (e.g., Naylor, 2000) and vegetation (Schad, 2000).

Fascinating topics, but outside the scope of this review. Here, we examine the importance of the Moon starting with a discussion of how the Moon holds a record of the early phases of the formation of the Earth–Moon system. On its surface, there are a great number of impact craters, whose features are essentially unmodified by internal or atmospheric processes. On Earth, the characteristics of the original impact craters are difficult if not impossible to discern. Actually, we cannot forget that the Moon has two distinctly different hemispheres. Its near side contains extensive dark lava flows (the "maria" in the words of Galileo Galilei) and impact craters of all sizes, from micrometers to very large dimensions, such as the crater Tycho, with bright ejecta extending for thousands of kilometers The far side has a much thicker, older crust, with more and deeper craters and fewer flat lunar maria than the near side. One of the largest craters in the Solar System, the Aitken basin near the lunar south pole, extending 2500 km and 12 km deep but practically invisible from Earth, is sited there.

The large number of impact craters on both hemispheres attest to the fact that in early times, the inner Solar System experienced several episodes of intense bombardments by bodies coming from its periphery. For clarity, Tables 3.1 and 3.2 provide the names of Earth's geological time intervals. Table 3.2 also gives an approximate correspondence with Martian geology.[1] From the longest to the shortest, these lengths of time are known as eons, eras, periods and ages.

For a review of the lunar surface chronology, see https://www.lpi.usra.edu/publications/books/lunar_stratigraphy/chapter_4.pdf.

Among the several bombardment episodes, the Late Heavy Bombardment (LHB) occurred approximately 4.1–3.8 Gy ago (see Bottke and Norman, 2017). On Earth, this period corresponds

TABLE 3.1   Name and Duration of Eons

Time (Gy)	Name	Duration (Gy)
4.6–0.6	Precambrian	4.0
0.6–0.07	Phanerozoic	0.53
0.07–0.008	Cenozoic	0.062
0.008–now	Quaternary	0.008

TABLE 3.2   Name and Duration of Eras

Time (Gy)	Name	Duration (Gy)	Approximate Correspondence with Periods on Martian Surface
4.6–4.0	Hadean	0.6	Pre-Noachian and Noachian
4.0–2.5	Archean	1.5	Hesperian
2.5–0.5	Proterozoic	1.5	
0.5–0.25	Paleozoic	0.25	
0.25–0.07	Mesozoic	0.20	Amazonian, which extends to today
0.07–0.03	Cenozoic	0.04	
0.03–now	Quaternary	0.03	

to the Neohadean and Neoarchean eras, namely quite early in Earth's history. During this interval, a large number of cosmic bodies collided with Earth, the Moon and the other terrestrial rocky planets, after they had already accreted most of their mass and formed a hard crust. The details of such bombardment episodes are still vigorously debated. For a recent analysis of the timeline of the lunar and Martian bombardment in the first billion years of Solar System history, see Morbidelli et al. (2018). In addition to the Moon and Mars, Ceres (the largest body in the Asteroid Main Belt and presently classified as the dwarf planet of the inner Solar System) gives useful information on the cratering history, as discussed by Strom et al. (2018).

An important question remains, namely how accurate is the dating of lunar soil using the samples brought back to Earth by the Apollo and Luna missions. Such a dating method seems not entirely conclusive, as discussed by Michael et al. (2018) summarizing the results derived from the potassium–argon (K–Ar) method (mostly based on the ^{40}Ar–^{39}Ar technique) of highlands rocks collected by those missions, and clasts dating of 23 lunar meteorites. In future studies, the dating of other basins, including the South Pole–Aitken, should be determined to further clarify the impact history. The above-cited papers are further proof of the great importance of future sample-return missions and the continued study of meteorites from the Moon, Mars and Vesta.

Small craters are produced on the lunar surface even today; see, for instance, Verani et al. (2001). Speyerer et al. (2016) discuss recent craters identified by comparing pairs of images taken by the Lunar Reconnaissance Orbiter (LRO). Figures from this research are shown in https://www.space.com/34372-new-moon-craters-appearing-faster-than-thought.html.

Apart from the many details regarding the bombardment episodes, a strong doubt remains: How could bacterial life survive on Earth? Well, it could. For instance, Abramov and Mojszis (2009 and later papers) modeled the effects of the LHB on Earth as a whole to assess the extent of habitable volumes on the crust for a possible near-surface and subsurface primordial microbial biosphere. The analysis showed that there was no plausible situation in which the Earth was fully sterilized. Furthermore, their results explain the persistence of microbial biospheres even on planetary bodies strongly reworked by impacts.

Indeed, life is incredibly resilient to adverse conditions; bacteria are found in extreme environments, such as volcanic ash, submarine thermal vents and even water from nuclear reactors. Small organisms, hibernated in Siberia 20,000 years ago, have been brought back to life. An amazing amount of microbial life has been found in boreholes as deep as 5 km, where the temperature is higher than 120°C and no light can reach the microorganisms (see the 2018 announcement on the Deep Carbon Observatory website: https://deepcarbon.net/).

3.3 THE PROTO-SUN AND THE EARLY EARTH

We previously mentioned the proto-Sun, also named the "young" Sun, which presents problems of its own. According to stellar evolutionary theories, the luminosity of the Sun in early times was about 80% of its present value, so that the surface temperature of Earth was below 0°C, too cold for liquid water, as pointed out by Carl Sagan around 1970. Therefore, strong heating mechanisms were needed to raise the Earth's temperature above freezing value. A powerful internal mechanism was plate tectonics, which enriched the atmosphere in CO_2 and to a lesser amount in CH_4. Methane and water vapor increased when bacteria started to flourish. Another powerful greenhouse gas could have been carbonyl sulfide (COS), produced by the many active volcanoes. Another source of heat was the decay of four isotopes affecting Earth's internal heat budget: ^{40}K (potassium), ^{235}U and ^{238}U (uranium) and ^{232}Th (thorium).

In addition to plate tectonics and all the other factors mentioned, we might conceive that the impacts of cosmic bodies helped to melt the thick ice crust of a "snowball" Earth. In other words, comets, asteroids and meteorites could have had an important role not only as deliverers of water and organic matter from the presolar nebula but also as powerful mechanical heaters.

The Sun itself may have been more effective at heating the Earth than estimated simply by its luminosity, because its solar wind and ultraviolet (UV) and X-radiation may have been much higher than at present. Indeed, strong winds of energetic particles and sudden flares of high-energy radiation are seen coming from the chromospheres of stars with surface temperatures ranging from approximately 2000 K to 4500 K (in astronomical jargon, stars of spectral types M and K, the Sun is a G-type star, with a current surface temperature of around 5800 K).[2] Additional information on the young Sun will be given in Chapter 7 when discussing meteorites.

Another important question is how effective was the early CO_2, the only greenhouse gas before bacterial life flourished and added methane as a second major component of the atmosphere. A delicate balance was needed to avoid, on the one hand, a runaway effect leading to a Venus-type condition and, on the other, its disappearance in favor of carbonates and silicates leading to a Mars-type situation. Thanks to plate tectonics, the replenishment cycle of CO_2 in the atmosphere, due to continuous modifications of the seafloor, had a typical duration of about 500 My, essential for the emergence and development of life. Comets and asteroids also had their part to play, because CO, CO_2 and H_2O ices are an important constituent of these bodies. The role of CO_2 as a thermostat has continued throughout the entire history of Earth, and will continue into the future. See, for instance, Cramwinckel et al. (2018).

The explosion of the human population and activities in the last centuries (the so-called Anthropocene) has had a large influence on the climate that overcomes the natural regulation mechanisms. If CO_2 and other greenhouse gases emissions, deforestation and soil and freshwater consumption continue at the present rate and the global temperature increases by 2°C or more above its current value, the planet might leave the glacial–interglacial cycle and enter into a new period of "hothouse Earth." See, for instance, Steffen et al. (2018).

In addition to CO_2, the role of oxygen was of paramount importance. Earth's oxygen levels rose and fell more than once before the planet-wide "Great Oxidation Event," which occurred about 2.4 Gy ago. Evidence that oxygen levels increased in distinct episodes before that great event comes from an examination of molybdenum and rhenium content in rocks, and nitrogen isotopes and selenium in marine sediments (Koehler et al., 2018). The data collected from nitrogen isotopes sample the surface of the ocean, while selenium is a proxy for oxygen in the atmosphere. A better determination of the evolution of the oxygen content in the atmosphere, from the time of the first photosynthesizing cyanobacteria perhaps as early as 3.5 Gy ago, to the 21% concentration of today, is important for a better understanding of the evolution of life on Earth. Oxygen levels during the Proterozoic eon, 2.5–0.5 Gy ago, are particularly important. See the reviews by Miller (2018) and Fitzgerald (2018), who give references to the original papers. When, more than 2 Gy ago, cyanobacteria started producing O_2 as a by-product of photosynthesis, the molecule was actually poisonous to most other species of microorganisms alive at the time, and mass extinction followed. The evolution of atmospheric oxygenation is thus critically associated with the evolution of life. The rise of O_2 to its present concentration was neither a single sudden jump nor a smooth and steady climb. Rather, the concentration increased in subsequent episodes separated by millions of years, and in between sometimes the concentration decreased. The exact evolution is not entirely understood, because geological records of the ancient atmosphere are rare. Geological and fossil records suggest that the formation of complex life was associated with oxygenation, but the details are still debated. Did rising oxygen levels in the oceans and atmosphere drive the formation of complex life? Were they unrelated or did the emergence of complex life cause a rise in oxygen? These questions, debated for instance in a series of 18 papers of a special issue of *Emerging Topics in Life Sciences* titled "Early Earth and the Rise of Complex Life" (Lyons et al., 2018) remain largely unanswered. In the frame of extraterrestrial life, it can be added that such studies are of great importance for evaluating oxygen concentration as a potential signature of life on other planets.

Notice that oxygen can be transferred from the Earth to the Moon, as evidenced by the JAXA Kaguya spacecraft data (Terada et al., 2017). For five days of each lunar orbit, the Moon is shielded from solar wind bombardment by the Earth's magnetosphere, which is filled with terrestrial ions (see, for instance, Barbieri et al., 2002, Wilson et al., 2006). Although the possibility of the presence of terrestrial nitrogen and noble gases in lunar soil has been discussed based on their isotopic composition, complicated oxygen isotope fractionation in lunar metals (particularly the provenance of a ^{16}O-poor component) remains an enigma. The paper by Terada et al. reports observations of significant numbers of 1–10 keV O^+ ions, seen only when the Moon is in the Earth's plasma sheet. Considering the penetration depth into metal of O^+ ions with such energy and the ^{16}O-poor mass-independent fractionation of the Earth's upper atmosphere, the authors conclude that biogenic terrestrial oxygen is transported to the Moon by the Earth's wind and implanted into the surface of the lunar regolith, at depths of around tens of nanometers. They also suggest the possibility that the composition of Earth's atmosphere billions of years ago may be preserved on the present-day lunar surface. The varying distance Earth–Moon, and the windows for habitability conditions on the Moon will be discussed in the following section.

3.4 ASTRONOMICAL EFFECTS ON TERRESTRIAL LIFE

In the previous section, we have seen that the balance of the different chemical species on land, lakes, seas and atmosphere is very delicate, and its study is very complicated. Conflicting data and theoretical models abound. Among the many factors that must be taken into consideration, let us mention astronomical effects.

A paper by Del Genio et al. (2018) examines the divergent climate and habitability histories of Venus, Earth and Mars, taking into account both geophysical and astronomical effects. These include processes that determined their interior dynamics and the presence or absence of a magnetic field; the surface–atmosphere exchange processes that acted as a source or sink for atmospheric mass and composition; the Sun–planet interactions that controlled the escape of gases to space; and the atmospheric processes that interacted with these to determine climate and habitability. The divergent evolutions of the three planets provide a valuable context for thinking about the search for life outside the Solar System.

In recent times, say the last 50 My or so over which computations are reliable, the Sun's luminosity stayed essentially constant, the position and extent of the continent were as today, and there was no large influence on Earth's climate by the infall of cosmic bodies.

However, variations in eccentricity and the inclination of the Earth's orbit and rotation axis obliquity came into play, with important effects on climate recognized by the Serb mathematician Milutin Milankovitch (main paper, Milankovitch [1930]; those cycles bear his name, Milankovitch cycles). Improved computer methods and ephemeris of Solar System objects have allowed reliable computations of the previous 50 million years (see Laskar et al., 2004a, 2011). Studies of climatic variations on Mars also provide interesting results (e.g., Laskar et al., 2004b). During the several million years before our epoch, the Earth cycled in and out of an ice age every 100,000 years or so. The planet left the last ice age around 12,000 years ago and is currently in an intergla-cial cycle called the "Holocene epoch." In this cycle, the Earth has natural regulatory systems that help keep it cool, even during the warmer interglacial periods.

On a very short timescale, an oscillation with a period of about 2,300 years (the so-called Hallstatt cycle) is found in cosmogenic radioisotopes (^{14}C and ^{10}Be) and in paleoclimate records through-out the Holocene. This oscillation is typically associated with solar variations, but its primary physical origin remains uncertain. Scafetta et al. (2016) show evidence of an astronomical origin for this cycle. Namely, this oscillation is coherent to a repeating pat-tern in the periodic revolution of the planets around the Sun: the major stable resonance involving the four Jovian planets (Jupiter, Saturn, Uranus and Neptune), which has a period of about 2318 years.

Over a much longer length of time, essentially from the very beginning 4.6 Gy ago, the Earth–Moon distance had consider-able variations. The Apollo and Lunokhod missions of the 1970s left six retroreflectors on the Moon that can be used to measure the distance by means of the echo of light pulses sent by power-ful ground-based lasers. According to such accurate laser rang-ing measurements, such as those performed at Grasse in France (Courde et al., 2017), at the Apache Point Observatory in New Mexico (https://www.apo.nmsu.edu/) and at the recent Chinese station (http://www.leonarddavid.com/china-uses-apollo-15-la-ser-ranging-hardware-on-the-moon/), the Earth–Moon distance increases by $\approx$4 cm/year. For fundamental mechanical reasons (the conservation of the total angular momentum), the variation of the distance entails a variation of the rotation rate of Earth. Future astronauts will have to traverse a longer journey, about 2 m, than the Apollo crews. Going backward at such a rate, a bil-lion years ago, a day lasted only 18 hours, there were almost 500 days in a year, tides were much more frequent and 10 times higher.

Moreover, the precession of the equinoxes was larger and faster, and the insolation conditions during the seasons were different than today. If such a decrease in the rate of distance stayed constant for 4 Gy, the Moon would eventually touch the Earth. However, extrapolation to the distant past is linear and very uncertain. A fundamental limit is encountered well before contact, namely the Roche limit (R_L) for disruption of the Moon by the Earth's tides:

$$R_L = 2.46 \text{ radius of Earth} \times (\text{density of Earth} / \text{density of Moon})^{1/2}$$

$$\approx 15,000 \ km \approx 1.2 \text{ Earth diameters}$$

Moreover, we cannot assume that at the very beginning neither the Earth nor the Moon was a truly rigid body, it is more likely that for quite some time each was a plastic deformable body. Even in more recent times, the separation speed was greatly affected by the extent of the continental lands, which changed several times, as discussed when examining plate tectonics. Geological evidence points to large variations in the duration of a day. The intricacies of the relationships between Earth and Moon in the earliest eras are exemplified for instance in a paper by Qin et al. (2018). According to this paper, the Moon's present tidal-rotational bulges are significantly larger than hydrostatic predictions, as already recognized by Laplace. The present bulges are likely relics of a former hydrostatic state when the Moon was closer to Earth and had larger bulges. Their formation was controlled by the relative timing of lithosphere thickening and lunar orbit recession, a geological process completed about 4 Gy ago when the Moon–Earth distance was less than $\approx$16 Earth diameters. In addition, the paper shows that in Hadean times the Earth was significantly less dissipative to lunar tides than during the last 4 Gy, possibly implying a frozen hydrosphere due to the fainter young Sun. Thus, the present recession speed seems an exceptionally large value.

Plate tectonics have been mentioned several times in Section 3.3. In 1596, the Dutch cartographer Abraham Ortelius suggested that the Americas, Eurasia and Africa had to be connected before the birth of the Atlantic Ocean. Such a suggestion was taken up by Benjamin Franklin and Alexander von Humboldt, but credit for the modern theory goes to Alfred Lothar Wegener (1966), who developed his ideas around 1912, among fierce controversy. Explorations of the magnetic properties of seafloors around 1960 led to the contemporary theory of plate tectonics. See Frankel (2012) for an exhaustive discussion of the continental drift controversy.

According to Wegener, the Pangea supercontinent existed some 250 My ago, in the mid to late Permian. Precambrian continents (e.g., Gondwana) may have existed as a united landmass (Rodinia) some 700 My ago. Recent simulations have been published spanning almost 1 Gy in the past. Two examples are shown in Figure 3.2, adapted from the UTIG PLATE project (see Lawver et al., 2009, http://www-udc.ig.utexas.edu/external/plates/recons.htm).

Moving to even earlier times, the trend of the lithosphere is very uncertain. Available geological data are severely limited for the first 2 Gy of Earth's history, due to the scarcity of relevant data and the non-uniqueness of interpretation. According to a review paper by Doglioni and Pansa (2015), the mass of the lithosphere grew continuously, with five, six or more episodes of continents joining together and then separating again. The general trend was a westward motion of the lithosphere, supporting the idea of a general tuning of the Earth's geodynamics and mantle convection by astronomical forces.

Another still unsolved major problem is the behavior of the magnetic fields of the Earth and Moon. The present structure of the terrestrial magnetic field shows an overall dipole pattern (at the present epoch tilted by some $11°$ with respect to the geographic poles, the North geomagnetic pole is located near Greenland), and several weaker multipoles. The situation was different in the past, and very dynamic, as indicated for instance by minerals such as magnetite. Other records are on chemical solutions later mineralized, or grains bound in deposits. The strength and even the polarity

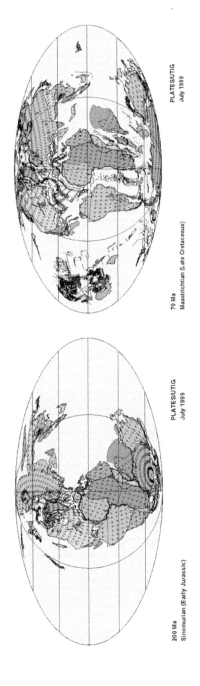

FIGURE 3.2   The disposition of continents 200 My ago (left). The situation at the time of the disappearance of the great reptiles, not much different from today (right). (Adapted from Plates/UTIG, Lawver et al., 2009.)

of the dipole had sudden changes and reversals, with no apparent periodicity. For a recent discussion based on the examination of a Chinese stalagmite radioisotopically dated at 107–91 thousand years before today, with an age precision of several decades, see Yu-Min Chou et al. (2018). One abrupt reversal transition has been detected occurring in only one to two centuries, when the field was weakest. These features indicate prolonged geodynamo instability.

The Moon probably possessed an initial dipolar field, not aligned with the rotation axis, and strongly coupled to that of the Earth. Such a primordial field went away when the body solidified and acquired its present synchronous rotation (see e.g., Oliveira and Wieczorek, 2017). However, the subject of the lunar magnetic field and its interaction with the terrestrial one is still open to debate. The so-called lunar swirls (see Figure 3.3) may provide additional information.

Most lunar swirls share their locations with localized magnetic fields. The bright-and-dark patterns may result when those magnetic fields deflect particles from the solar wind and cause some parts of the lunar surface to weather more slowly. According to

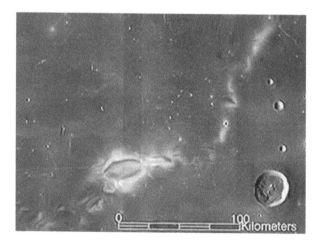

FIGURE 3.3 The Reiner "gamma-shaped" lunar swirl. (Image credit by NASA.)

a paper by Hemingway and Tikoo (2018), each swirl must stand above a magnetic object buried close to the Moon's surface. The picture is consistent with lava tubes (long, narrow structures formed by flowing lava during volcanic eruptions) or lava dikes (vertical sheets of magma injected into the lunar crust). Future lunar explorations might well study those swirls from in situ investigations.

In summary, although more work is needed to clarify the complex series of Earth–Moon interactions, in the earliest times land and sea tides on Earth must have been more frequent and higher than at present (in a first approximation, tide amplitude goes with $1/(distance)^3$, energy dissipation with $1/(distance)^6$). Equally larger and faster was the precession of the rotation axis, which determines the duration of the seasons and insolation conditions.

Do current living organisms, in particular the bacteria populating the Earth since the very beginning, keep a memory of such a dynamic and magnetic evolution of the Earth–Moon system? Probably not, they adapt very quickly to the changing environment.

Another debated problem is the origin of the Earth–Moon (E–M) double planet. According to several lines of evidence, the Moon originated from the giant impact of a Mars-size body with the Earth. Such a canonical model is able to explain the large iron deficit in the lunar mantle and the rapid initial spin of Earth (Canup, 2004). A crucial problem with the canonical model is the essentially identical isotopic composition in non-volatile elements (O, Ti, Cr, Si, W), in particular the percentage of the three stable isotopes of O. Several alternative formation hypotheses have been formulated, for instance involving multiple impacts of sub-Mars-size bodies creating several moonlets that merged together. Lock et al. (2018) invoke a much more energetic, high angular momentum impact that drove the Earth into a fast-spinning, vaporized state extending for tens of thousands of kilometers. Such a donut-shaped object is called a "synestia." As this material cooled, it condensed and formed the Moon, surrounded by Earth-composition vapor at pressures of tens of bars. The lunar orbit within such vaporized material lasted long enough to chemically equilibrate with the vaporized Earth, thus explaining the similarity of the isotopic and chemical compositions of the Moon and Earth. To summarize, explaining the compositional similarities of E–M mantles is extremely difficult. Single impact models require additional processes to erase compositional differences and/or alter the angular momentum.

From the previous considerations, we concluded that the earliest date of the appearance of life on Earth was from approximately 4.0 to 3.8 Gy ago. A comprehensive discussion of astronomical, geophysical and biological evidence (Pearce et al., 2018b) puts constraints on the likely windows; the habitability boundary could be as early as 4.5 Ga, the earliest possible estimate of the time at which Earth had a stable crust and hydrosphere, or as late as 3.7 Ga, the largest uncertainty coming from astronomical data.

Before leaving this chapter, it is worth reconsidering where we are in geological time. Currently, we are in the Phanerozoic eon, Cenozoic era, Quaternary period, Holocene epoch and upper Holocene age. The subdivisions become finer and finer with the increasing amount of information in more recent times. According to a proposal by the International Commission on Stratigraphy (ICS), the upper Holocene could be renamed Meghalayan. The Meghalayan is one of three newly named ages. The other two ages are the Greenlandian (11,700–8,326 years ago) and the Northgrippian (8,326–4,250 years ago). The beginning of each age is determined by the chemical signatures found in rock samples; each signature relates to a big climatic event. The Greenlandian, the oldest age of the Holocene (also known as the "lower Holocene"), began 11,700 years ago, as Earth left the last ice age. The Northgrippian (also known as the "middle Holocene") began 8,300 years ago when Earth abruptly began cooling. The Meghalayan (also called the "upper Holocene") started 4,250 years ago when a mega-drought devastated civilizations across the world. However, such a fine division of the Holocene is not generally accepted; see, for instance, the debate in https://www. sciencemagazinedigital.org/sciencemagazine/10_august_2018/ MobilePagedArticle.action?articleId=1416235&app=false#articl eId1416235.

NOTES

1. Notice that a different chronology and denomination of epochs on Mars has been proposed by Bibring et al. (2006), based on the data of the OMEGA experiment on board ESA's Mars Express.

2. The original Yerkes spectroscopic classification (also known as the Morgan and Keenan classification, MK) of stars according to decreasing surface temperature uses seven capital letters, O, B, A, F, G, K, M, with a finer subdivision expressed in numerals, e.g., G2 for the Sun, A0 for Vega. The coolest stars, with temperatures below 2000 K, are designated with the letters L and T. A Roman numeral, such as I, II,…, V is used to indicate the luminosity, from supergiants I to solar type V. The complete classification for the Sun is then G2-V.

CHAPTER 4

Main Characteristics of Living Organisms

F ROM THE PREVIOUS CHAPTERS, we derived that approximately 4.3 Gy ago, our Earth and Moon were already formed and solid. After less than 500 My from that beginning, a thick atmosphere, vast oceans of salty water, geothermal ponds and lakes, the first complex molecules, amino acids, proteins, RNA, DNA and the first living organism appeared. Bacteria dominated, and still dominate, life on Earth. It was only approximately 800 My ago that living organisms assumed macro dimensions and the capability to move around and forage for food. Great flying animals and reptiles disappeared about 66 My ago. Mammalians and finally humans were the end-product of such an evolution. Therefore, evolution seems to augment the complexity of living organisms, a sort of thermodynamic and philosophical contradiction, as already recognized by Schrödinger. Why does life seem to organize itself in entities of continuously increasing complexity and order, while in the Universe as a whole, disorder and entropy augment? The immediate explanation is that life on Earth is not a closed system. Moreover, even if present on many planets, life may represent a minuscule insignificant accident in the evolution of the Universe. We cannot ignore the end point of life, namely death, in such considerations. Living cells and organisms are programmed to die. No doubt, death is a necessary connotation of life, at least of life based on chemical and biological processes. Similarly, does our technological civilization, the end point of a continuous augmentation of complexity, contain the germs of its destruction? A limited time span of technologic societies might act as a guillotine to our capability to discover other advanced civilizations, if self-destruction is a universal connotation of such civilizations. Several authors debate such questions. Among many books, the one written by two of the most important astrophysicists of the last decades, R.M. Bonnet and L. Woltjer (2008), titled *Surviving 1,000 Centuries: Can We Do It?*, deserves a mention. The book cover depicts a comet lingering above a scene of destruction, further attesting to the importance of cosmic bodies for life and death. Humankind, at any rate, is today a most important player in the evolution of Earth. As stated in Chapter 3, according to several authors the present epoch should be named "Anthropocene" to underline such importance. Where Bonnet and Woltjer consider a long time span, other scientists are much more concerned, e.g., the late Stephen Hawking, who always promoted the rapid colonization of Mars, say in a few centuries, to make humanity survive. The colonization of the Moon and Mars will be discussed in Chapter 6.

As already mentioned in Chapter 2, the chemical composition of living beings is dominated by four basic elements: H, C, O and N. With them, 20 fundamental biotic amino acids are built, all with left-handed symmetry (chirality). Moreover, life needs liquid water and a source of energy for metabolism. Notice also that mammals (including humans) apparently use only 24 chemical elements, the "big four" elements being responsible for more than 90% of their weight.

Could the cosmic origin of life be more effective than the simple provision of chemical elements? Could basic ingredients such as amino acids come from the outside, born in interstellar clouds and carried to Earth by comets, asteroids and meteorites? In the opinion of the writer, yes, at least in part. Such a hypothesis has the advantage of giving much more time to produce complex molecules than the 4 Gy provided by Earth. Even the left-handed symmetry of biotic amino acids might have a cosmic origin. Some line of evidence supporting this cosmic hypothesis will be given here and later on when speaking about comets in Chapters 7 and 8.

As already stated, amino acids are simple organic compounds made of H, C, O and N; in a few cases, sulfur (S) is also present. The ($-NH_2$) is the amino group and the ($-COOH$) is the carboxyl group. Only 20 amino acids (usually called α amino acids) are common in humans and animals, with two additional amino acids present in a few species. Over 100 amino acids are found in other living organisms, particularly plants. Meteorites, too, carry a large complement of amino acids, often not found in terrestrial material. Twenty amino acids strung together in a chain become a protein. A chain of just 200 amino acids provides the potential for some 10^{260} proteins. Chains as long as 2000 amino acids are known. The order of amino acids in the chain is extremely important as not all combinations of amino acids make functioning proteins. Actually, no matter what the sequence of amino acids, the main number of shapes is limited to four (see e.g., Cockell, 2017). In Chapter 8, we will see that glycine, the simplest of amino acids, NH_2-CH_2-COOH, has been found on the atmosphere of comet 67P C-G using the instrument Rosina aboard the European Space Agency's (ESA) cometary mission Rosetta.

4.1 THE PUZZLE OF CHIRALITY

As already mentioned, a peculiar characteristic of all biotic amino acids except glycine is a left-handed spatial arrangement (chirality), while sugars (including ribose, the backbone of DNA) are right-handed. In chemistry, one variety is called an "enantiomer." Amino acids obtained by synthesis in the laboratory are instead a mixture of both arrangements (a racemic compound).

Such chirality of biotic amino acids is still unexplained, and several scientists attribute it to cosmic causes (see e.g., the book by Guijarro and Yurs (2007) titled *The Origin of Chirality in the Molecules of Life*). A readable account of chirality and its possible origin in space is given in https://arstechnica.com/science/2018/02/how-did-life-begin-its-chemistry-101-but-in-space/.

Several attempts have been made to reproduce the conditions of such cosmic origin. For instance, a group of researchers at Hokkaido University in Japan performed a laboratory experiment at the very low temperature of 12 K typical of interstellar nebulae. They succeeded in substituting an H atom with a D atom. At this point, not only did the glycine become chiral, but it also acted as a catalyzer of chirality of other molecules (Oba et al., 2015). Another example of the possible cosmic origin of chirality is provided by an excess of left-handed structures in the α-hydrogen aspartic and glutamic amino acid in three fragments of a C-2-type carbonaceous chondrite meteorite found in Tagish Lake in British Columbia, Canada (Glavin et al., 2012). The large meteoroid exploded in January 2000, raining fragments across the frozen surface of the lake. Because many people witnessed the fireball, pieces were collected within days and preserved in their frozen state. This ensured that there was little contamination from terrestrial life. Aspartic acid is an amino acid used in every enzyme in the human body. The amino acids of the meteorite were probably created in space, as revealed by isotopic analysis, in particular by the abundance of ^{13}C vs ^{12}C. Since the chemistry of life prefers lighter isotopes, amino acids enriched in the heavier ^{13}C were likely created by non-biological processes in the parent asteroid. Interestingly, another amino acid found in the fragments, alanine, showed the same abundance of ^{13}C but not the left-handed excess. The authors conclude that the ^{13}C enrichment combined with the large left-handed excess in aspartic acid but not in alanine provides very strong evidence that some left-handed amino acids used by life to make proteins can be produced in excess in asteroids. More difficult is to find the process responsible for such excess. Exposure to polarized radiation in the solar nebula has been invoked in other cases. However, in the case of Tagish Lake the excess is too large, and is present only in aspartic acid and not in alanine. A possible explanation lies in the different crystallization shapes of the two amino acids. A small initial excess in the aspartic acid may have been amplified by dissolution in water inside the asteroid.

We have described this paper at some length because the results, if confirmed by further data, may complicate the search for extra-terrestrial life: since a non-biological process can create a left-handed excess in some kinds of amino acids, such an excess alone doesn't constitute solid proof of biological activity.

Chirality in cosmic sources can indeed precede life on Earth and most likely the formation of the Solar System. McGuire et al. (2016) discovered the first chiral molecule (propylene oxide, CH_3CHOCH_2) in an interstellar cloud in the direction of Sagittarius B, toward the center of the Milky Way. The observations were first made with the Green Bank Telescope Prebiotic Interstellar Molecular Survey (PRIMOS) and then with the Parkes radiotelescope in Australia. The molecule appears as an absorption feature in the spectrum of a cold and extended molecular shell. Such a shell surrounds the massive protostellar clusters in the Sagittarius B2 star-forming region. This work raises the prospect of measuring the enantiomer excess in various astronomical objects, including regions where planets are being formed, to discover how and why the excess first appeared.

4.2 LIVING ORGANISMS

Let us examine again some characteristics of life: When can an assembly of molecules be defined as a "living organism"? Large efforts are underway to answer such a difficult question: the frontier separating living from non-living beings is surely not a well-defined one, and a source of wide debate with respect to extra-terrestrial life.

A very generic definition could be the following: a living organism is a molecular chain capable of

1. Self-organizing in tridimensional structures.
2. Utilizing energy sources to perform chemical reactions (metabolism).
3. Storing, copying and utilizing biochemical instructions for reproduction.
4. Evolving through adaptation, mutations, error corrections, repair, adapting to the changing environmental condition, through natural selection.

These and other considerations have led to the hypothesis that a still not well understood chain of processes gave origin to a Last Universal Common Ancestor (LUCA), which lived about 3.9 billion years ago and whose genetic traces are present in all living beings, including ourselves. We must recall that in 1859, Charles Darwin, in his seminal book *On the Origin of Species*, suggested a common origin of all living beings through the evolutionary process. To be sure, LUCA was not the origin nor the first living being, it already had a complex life form with many mechanisms to convert information between DNA, RNA and proteins. In essence, LUCA was simply the last universal, only one of many early organisms of which all but one died out.

Where did life begin on Earth? According to several theories, prior to the origin of simple cellular life, the building blocks of RNA (nucleotides) had to form and polymerize in favorable environments on the early Earth. At this very early time, celestial bodies delivered organics such as nucleobases to geothermal ponds, whose wet–dry cycles promoted rapid polymerization. RNA polymers emerged very quickly after such deposition of cosmic origin, and the synthesis of nucleotides and their polymerization into RNA occurred in just one to a few wet–dry cycles, greatly aided by much larger and more frequent tides. Under these conditions, RNA polymers likely appeared prior to 4 Gy ago (see, for instance, Pearce et al., 2018a). The title of this paper contains the expression "RNA world," a hypothesis about the origin of life. However, according to Kim et al. (2018), the early RNA may have been different from the modern form, made of four main building blocks: nucleobases called adenine (A), cytosine (C), guanine (G) and uracil (U), together with a backbone of sugar and phosphate. The authors propose that inosine could have served as a surrogate for guanosine in the early emergence of life.

Taking into account the previous considerations, we can conclude that the often repeated statement "we only know of one life, ours," is a bit deceiving: in four billion years the Earth went through a series of different forms of life and quite possibly we do not even know all of them. The increasing knowledge of earthly life will greatly help us to recognize different forms of life on the planets of other stars.

CHAPTER **5**

Water and Life

HOWEVER COMPLEX AND INTRICATE the past history of Earth and the appearance and development of life on it, water appears as a fundamental ingredient, reminding us of the prescientific ideas of Thales of Miletus (624–546 BC), the great engineer, mathematician and natural philosopher: water is the principle of everything (*arké*, in Greek ἄρχειν).

On the origin of earthly water, there are two main theories:

- The "endogenous" theory: Water originated in the deep interior of the Earth, and was brought to the surface by processes such as volcanism and plate tectonics.

- The "exogenous" theory: Water came from the outside, carried by the infall of planetesimals, comets, asteroids and meteorites.

Probably both mechanisms were at play, but let us first examine how much water there is on Earth. In addition to the estimated amount of several oceans of water in the deep interior (see the already quoted paper by Doglioni and Panza), a recent line of evidence (Nestola et al., 2016, Spiga et al., 2019) adds other water in the region of 500–600 km deep. This finding comes from an examination of diamonds found in a Brazilian mine. The ring-woodite mineral included in such diamonds contains about 1.4% of OH⁻ ions. The hydrated minerals in such a deep region contain the equivalent of two to three oceans. Such a large amount of water above the lower mantle of the Earth might have come from the continuous accretion of planetesimals in the early days of the formation of Earth. Thus, the discovery of such deep water adds further interest to the study of the role of comets, asteroids and meteorites in connection with terrestrial water.

5.1 SOME PROPERTIES OF WATER OF ASTRONOMICAL INTEREST

The peculiar characteristics of water as a solvent are of paramount importance for life. In passing, we quote that alternative forms of life based on different solvents, such as ammonia (NH_3) or methyl alcohol (CH_3OH), both present in interstellar space, are considered in scientific papers and science-fiction novels. Here, we concentrate on those water properties relevant to astronomical detection and studies.

1. The temperature–pressure phase diagram. Water can be found as a gas, liquid or solid according to the values of pressure and temperature. On the surfaces of many celestial bodies, from the Moon to present Mars to asteroids and to comets, conditions are such that no liquid water can exist, at least not on large areas nor for a sizeable amount of time. An immediate consequence of the phase diagram is the concept of a habitable zone (HZ) around a particular star. A necessary condition for life, at least as we know it, is the right distance from the star to maintain water in the liquid phase. In the current epoch, Venus is too hot, Mars is too cold, only Earth is inside the ideal zone. This situation does not necessarily apply to the early phases of the Solar System. The already mentioned paper by Del Genio et al. (2018) considers the possibility that all three planets might have simultaneously been habitable early in their histories.

In general, the extent of the HZ depends on the star, namely on its mass, radius and surface temperature. In addition to the presence of liquid water on extended regions of the rocky exoplanet surface, the classical HZ assumes that the most important greenhouse gases are CO_2 and H_2O. Recent HZ reformulations enlarge the classical zone, pointing to the possible diversity of habitable exoplanets. Ramirez (2018) reviews the planetary and stellar processes considered in both classical and extended HZ formulations.

A word of caution here: liquid water may be necessary, but surely not sufficient for habitability. Unfortunately, the media give the impression that "habitable" means tout court "inhabited." Not correct and confusing. More conditions for true "habitability" must be met:

a. A thick atmosphere protecting from sterilizing particle and electromagnetic radiations from the star and other cosmic sources and from the impacts of cosmic bodies. Large-scale winds have the important function of thermalizing the planet, as liquid oceans do, even if the planet shows the same face to its star.

b. An ordered magnetic field protecting from cosmic rays.

c. A massive moon stabilizing the rotation axis and the duration of the seasons.

d. A single central star with constant luminosity, ensuring both a stable orbit to the planet and constant insolation for very long times.

For instance:

• Mars violates Conditions 1, 2 and 3, as discussed later on.

• Many extra-solar planets violate Condition 4. Our Sun is a remarkable example of a single stable star. Many stars instead are in binary or even more complex systems, and display luminosity variability of high amplitude, often with X-ray flares.

Therefore, life elsewhere may not be as common as usually believed. See, for instance, Ward and Brownlee (2009) and Livio (2018).

2. Water can be in ortho or para states according to the spin orientation of the two H (parallel in ortho, anti-parallel in para). The ortho vs para (OPR) ratio can be measured with suitable radiotelescopes such as *Institut de Radioastronomie Millimétrique* (IRAM; described in Chapter 9). The OPR is sensitive to the formation temperature of the molecule. Moreover, in the "para" status, the molecule does not interact with the magnetic fields pervading the interstellar medium.

3. Detectability by radar. Radars, either on ground stations or on spacecraft, can detect water vapor in the atmosphere, iced and liquid water on surfaces and even underground. Ciarletti (2016) gives a review of radars for Solar System studies. The presence of water in celestial bodies can be confirmed by magnetometers onboard satellites and by navigational data providing the gravity field and internal structure of the celestial body. Examples will be given in the following chapters.

4. Different isotopic composition. Deuterium (D) is the stable hydrogen isotope of astrophysical interest. The percentages of H_2 vs HD vs D_2 can be measured on celestial bodies using different techniques, as can the percentages of ^{16}O vs ^{17}O vs ^{18}O. For instance, the Atacama Large Millimeter/Submillimeter Array (ALMA) telescope in Chile (described in Chapter 7), using its band from 0.3 to 0.4 mm (787–950 GHz) discovered a jet made of hydrogen, deuterium and oxygen (HDO) flowing away from a massive star in the nebula NGC 6331 (see McGuire et al., 2018). This most important topic of the isotopic composition of water will be discussed in reference to the Rosetta mission to comet 67P in Chapter 8.

5. Density lowers when freezing. Water is one of the rare substances that decreases its density when freezing. Therefore, ice rises to the surface of the liquid, and in so doing protects the life present down under. Moreover, the salinity of ice is lower than that of deep water, and ice is often the purest phase of water. The increase in salinity with depth has another consequence, important not only for Earth but also for Mars, Europa and Enceladus, namely that water remains liquid even several degrees below $0\,°C$. The presence of underground water in these bodies will be discussed in Section 5.3.

Actually, the properties of water under the peculiar conditions of the cosmic ambient are not entirely known. For instance, ultraviolet (UV)-irradiated amorphous ice behaves like a sticky liquid at low temperatures. According to Tachibana et al. (2017), at temperatures between $-210\,°C$ and $-120\,°C$ this liquid-like, high-viscosity ice may enhance the formation of organic compounds including prebiotic molecules and the accretion of dust to form planets. As a second example of the peculiar properties of water, a new form of super-ionic ice, which could be present in the mantles of Neptune and Uranus, has been found by Millot et al. (2018).

5.2 PROXIES OF WATER

In several astronomical situations, water cannot be detected directly. Proxies are instead used to infer its past or contemporary presence, such as the proton flux from the surface, the molecule H_2 and the negative ion OH^- included in hydrated minerals. Notice that our atmosphere contains the neutral radical OH, coming from the dissociation of ozone O_3 by solar UV photons. The function of the ozone layer approximately 30 km above ground as a powerful protection against such sterilizing radiation is well known. The atmospheric OH itself is important in the present context in a second way, because it has a detergent action on methane CH_4 (an indicator of bacterial life), leading after a complex chain of chemical reactions to gaseous CO_2 and H_2O.

Coming back to OH⁻ in hydrated material, we can quote the map of the lunar surface provided by the Indian spacecraft Chandrayan-1 (see https://www.isro.gov.in/Spacecraft/chandrayaan-1).

Measurements by the NASA spacecraft Lunar Reconnaissance Orbiter (LRO, https://lunar.gsfc.nasa.gov/) have shown that the proton flux from the lunar surface is due to the interaction of cosmic rays with hydrated material stored in the regolith. In addition, data from space missions, ground radar results, as well as sensitive analyses of lunar rock and soil samples brought back by the Apollo missions or stored in lunar meteorites, indicate that the Moon is not as dry as was thought until a few years ago. In addition to the occurrence of water ice in permanently shadowed polar craters, the presence of hydrated materials at high, not shadowed latitudes is now certain. Moreover, studies of the products of lunar volcanism indicate that the lunar interior also contains more water than was once appreciated. The lunar mantle may even be as water-rich as Earth's upper mantle. The existence of indigenous sources of water implies that the Moon may not always have been as dead and dry as it is today. Insofar as water is required for habitability, one can speculatively identify two possible windows for lunar habitability. The first may have occurred immediately following the accretion of the Moon, the second some hundreds of millions of years later following outgassing associated with lunar volcanic activity (Schulze-Makuch and Crawford, 2018).

To close this section, we quote that a flux of protons has been measured from the surface of Ceres by the NASA DAWN spacecraft, indicating the presence of water ices on the surface of the dwarf planet.

5.3 UNDERGROUND LIQUID WATER IN THE SOLAR SYSTEM

Enormous amounts of water vapor are found not only in the interstellar and even intergalactic medium, but also in the atmospheres of stars (including the sunspots of our Sun), and on planets of "our" and "their" solar systems.

The same is true for water ices, found everywhere, adsorbed on interstellar dust grains and buried in crevices and craters on the Moon, even on Mercury, on comets and asteroids.

What about liquid water? Is it present only on Earth? No, underground liquid water exists on Mars, and it is surely present under the icy crusts of several Jupiter and Saturn moons, of Pluto and of Ceres. It is too early to be certain about liquid water on extra-solar planets, but knowledge will advance rapidly and confirm the likely suspicion.

Space missions have conclusively shown the presence of subsurface water on

- Mars, the most obvious target for detecting alien life. The detection of underground possibly salted water by the MARSIS radar on board the European Space Agency's (ESA) MarsExpress will be discussed in Chapter 6.

- Europa, Ganymede, Callisto, moons of Jupiter.

- Titan and Enceladus, moons of Saturn.

- Pluto, formerly the ninth planet, and since 2006 the prototype of the new class of dwarf planets (still today a much debated decision; see, for instance, Metger et al., 2018).

- Ceres, formerly the first asteroid, and presently the dwarf planet of the inner Solar System.

Regarding Jupiter's moons, beautiful images of the Medicean moons discovered by Galileo Galilei in Padova in January 1610, were obtained in 1997 by the NASA Galileo spacecraft. Europa, the smallest of the four moons, displays a fractured icy crust. Under such crust, some 20 km thick, a liquid ocean of water was suspected to exist above a rocky nucleus, as inferred from navigational and magnetometric data. Twenty years after the Galileo mission, the Hubble Space Telescope hinted at a geyser of water vapor coming out of the surface, adding evidence to the existence of an underground liquid ocean (Sparks et al., 2017). On the basis of such an exciting possibility, the old Galileo spacecraft data have been reanalyzed, and the existence of gaseous ionized plumes has been confirmed (Jia et al., 2018).

How does water remain liquid at extremely low surface temperatures? Most likely, the warm liquid water contains plenty of salts, due to the excellent capability of water to dissolve the rocks in the underlying nucleus of the moon. High salinity helps but is not sufficient. The strong tide exerted by Jupiter on the crust of Europa and the ensuing friction provide the needed energy.

An ocean of water may also exist under the crust of Ganymede, the largest moon of the Solar System with a magnetic field of its own. From an examination of the Galileo magnetometer data and aurorae discovered by the Hubble Space Telescope, the presence of a subsurface liquid (not necessarily salted) water ocean can be inferred (Saur et al., 2015). The amount of liquid is more or less the same as in the Earth's oceans. It may be that even Callisto, the outer Medicean moon, harbors such an ocean.

Two space missions are planned for the near future to find out more: JUICE (http://sci.esa.int/juice/), a joint ESA–NASA mission, and NASA Europa Clipper (https://europa.nasa.gov/internal_resources/125). Their launch could happen around 2022.

Regarding Saturn, the Cassini–Huygens mission has been and still is of paramount importance. Cassini–Huygens, one of the most spectacular missions ever performed, a joint enterprise among NASA, ESA and the Italian Space Agency (ASI), ended its life on September 15, 2017, by plunging into the planet's atmosphere. One if its first results was the exploration of Titan, the second-largest moon of the Solar System, where the Huygens module landed in 2005. In addition to the rain of methane and hydrocarbon and liquid lakes on its surface (see Figure 5.1), a subsurface liquid water was inferred. The formation of such a buried ocean and its influence on the primitive atmosphere of the moon

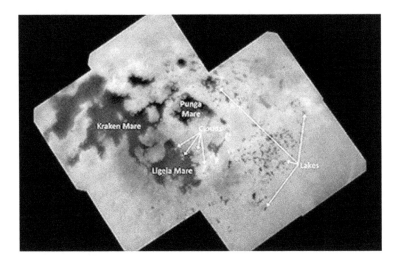

FIGURE 5.1   During Cassini's encounter with Titan, the spacecraft captured this infrared view of the moon's north polar lakes and seas, which are filled with liquid methane and ethane. To set the scale of the image, Kraken Mare is 1200 km wide. Small clouds are also visible. (Image credit: NASA/JPL-Caltech/SSI).

are discussed by Marounina et al. (2018). The authors show that the relative abundances of CO_2 and NH_3 determine the composition of Titan's atmosphere. Furthermore, the paper investigates the conditions for the formation of CH_4-rich clathrates hydrates at Titan's surface that could be the main reservoir of methane for the present-day atmosphere.

The importance of clathrate hydrates as protecting cages for volatile molecules such as methane will be examined Chapter 7 due to their relevance for comets. It is worth recalling that Titan's atmosphere contains molecules of monodeuterated methane (CH_3D) with a relative abundance D/H of 1.65 (+1.65, −0.8) × 10^{-4}, nominally eight times higher than the most commonly accepted value for the protosolar value D/H = 2 × 10^{-5} (de Bergh et al., 1988). The value found on Titan for D/H in methane is comparable to the D/H ratio measured in terrestrial H_2O, as discussed further in Chapters 7 and 8.

When Cassini examined the moon Enceladus, which has a diameter of 500 km and displays a peculiar fractured surface (see Figure 5.2), a geyser 30 km high above the surface was discovered in the South Pole region (see Figure 5.3). Such a geyser most likely originates from a warm underground ocean, and vapors find their way through cracks in the surface ice.

The presence of a subsurface ocean was confirmed by the onboard radar. Examining the net of surficial cracks, Lucchetti et al. (2017) found that near the South Pole the fracture penetration is about 30 km, likely reaching the ocean–ice interface. In the other regions analyzed in the paper, the depth of fracture penetration increases from 30 to 70 km from the South Pole to northern regions up to 75° latitude.

An initial analysis of the plume demonstrated that almost 98% of the geyser vapors is water; about 1% is molecular hydrogen and the rest is a mixture of CO_2, CH_4 and NH_3. A reanalysis

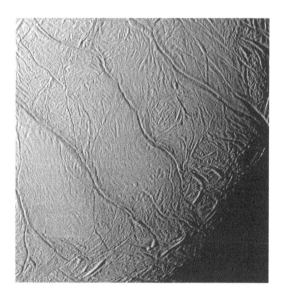

FIGURE 5.2 The system of fractures on the surface of Enceladus. The ground resolution is about 120 m. (Image credit: NASA/JPL/Space Science Institute).

FIGURE 5.3   An image of Enceladus plumes. (Image credit: NASA/JPL/ Space Science Institute).

of the gases (Postberg et al., 2018) has allowed the identification of fragments of large, complex organic molecules with molecular masses above 200. The presence of large complex molecules comprising hundreds of atoms is very rare beyond Earth and, along with liquid water and hydrothermal activity, bolsters the hypothesis that the ocean of Enceladus may be a habitable environment for life (see also https://www.jpl.nasa.gov/news/ news.php?feature=7174&utm_source=iContact&utm_medium= email&utm_campaign=NASAJPL&utm_content=daily 20180627-2).

Barge and White (2017) performed experiments to test the connection between hydrothermal vents and life on Enceladus and other worlds. Actually, Enceladus provides an ideal test case, since the pressure at the ocean floor is more easily simulated in the laboratory.

Let us now examine Pluto and Charon. Since 2006, Pluto is considered the prototype of dwarf planets found in great numbers beyond the orbit of Neptune. The New Horizons NASA mission has explored the system of Pluto in detail. A beautiful image of the

Tenzing Montes peaks (Figure 5.4) reveals that these mountains are highest on Pluto, ranging from 3 to 6 km above the surface.

Such height adds confidence to the suspicion that the surface is made of water ice. Methane ice, for example, would not be strong enough to hold up features that large. In addition, the orbital resonance between Pluto and Charon suggests the presence of a liquid ocean under the crust of Pluto. See the papers by Schenk et al. (2018a,b).

Regarding Ceres, the dwarf planet of the inner Solar System, the NASA DAWN mission obtained beautiful images and data, leading to the suspicion of a liquid water ocean below a thin crust. Such crust is composed of a mixture of ice, salts and hydrated materials subjected to past and possibly recent geologic activity. Marchi et al. (2018) have performed a detailed examination of the carbon-rich and aqueously altered Ceres surface, pointing out the possible influence of asteroid bombardment, as happened also in the cases of the Earth, Moon and Mars. Surficial water ice has

FIGURE 5.4 An image of the Tenzing Montes peaks on Pluto created with New Horizons data released on July 10, 2018. (Credit: Paul Schenk/ Lunar and Planetary Institute).

been detected in deep, permanently shadowed craters. The gravity data suggest that there is a softer, easily deformable layer beneath Ceres' rigid surface crust, which could be the signature of residual liquid left over from a past ocean.

By considering all the evidence accumulated in the foregoing cases, the fascinating hypothesis can be made that under the crust of Mars, of the moons of Jupiter and Saturn, of Pluto and Ceres, conditions exist that are favorable to bacterial life. Serious and imaginative scientists go to the extreme of supposing that civilizations may exist in such hidden oceans, forever secluded from the rest of the universe, unless we go down there and communicate with them.

CHAPTER 6

Human Outposts on the Moon and on Mars

FINDING LIQUID WATER UNDER the surfaces of the moons of Jupiter and Saturn, of Ceres and Pluto, is surely quite interesting, scientifically speaking, but the presence of liquid water under the soil of Mars and Ceres, or water ices on the Moon or on near Earth asteroids, has important practical meanings for our life up there. At present, optimism is widespread both in national space agencies and in private companies. However, we cannot forget how difficult it is to export human life elsewhere, especially for long durations or even permanent outposts. Crew must be able not only to arrive there but also to come back home and have their life protected in such harsh environments. Cosmic rays provide an example of perils having a cosmic origin. Schwadron et al. (2018) show that the dangerous radiation from deep space is intensifying faster than previously predicted. Data from the Cosmic Ray Telescope for the Effects of Radiation (CRaTER) aboard NASA's LRO show that cosmic rays in the Earth–Moon system are peaking at levels never seen before in the Space Age. The worsening radiation environment is a potential peril to astronauts, curtailing how long they could safely travel through space. The number of days a 30-year-old male astronaut flying in a spaceship with 10 g/cm^2 of aluminum shielding could go before hitting NASA-mandated radiation limits was 1,000 days in 1990, only 700 days in 2014. Surely, the many efforts under way will invent better shields but cosmic rays are only one of the risk factors of space travel. Other well-known problems are the loss of calcium from the bones, loss of muscular tone, permanent damage to the vision, psychological disturbances and so on. Long duration stays of astronauts in the ISS allow the test of remedies.

Adding all other ingredients, including innovative technologies, cost and political determination, it is plainly evident that carrying out the colonization of outer space is a big enterprise. The date of the next human landing on the Moon after the Apollo mission is not determined yet, not to speak of landing on Mars.

No matter how great the challenges and risks are, the motivations are also compelling, from sheer scientific interest to industrial, commercial and touristic enterprises, or, in a pessimistic attitude, simply to save humanity from extinction on our doomed planet. Building such outposts might cancel past geological records, for instance by careless excavation or mining. Carrying aboard the ship unwanted bacteria from Earth may endanger possibly existing exo-life, as was done in past centuries by colonizing Africa, Hawaii and Australia. We might even do such contamination on purpose, e.g. to make conditions on Mars more favorable to human colonies, an undertaking often called terraforming.

There is another aspect of human colonization that requires co-ordinated efforts, namely the protection of Earth against contamination by materials brought back from alien worlds.

These caveats and preventive measures to mitigate their effects are included in current plans so that we can be confident that all efforts will be made to protect the records of the history of the celestial body, alien life, our life there and here on Earth.

Let us consider now how the relevant players conceive their manned bases on the Moon and on Mars, although at the time of writing no plan is definitely consolidated and we should not be surprised by drastic changes to current projects before the first human launch takes place.

6.1 THE MOON

We have already pointed out the presence of hydrated materials and water ices in several places of the lunar surface, with an abundance much greater than expected a few years ago. See for instance Bandfield et al. (2018).

Figure 6.1 shows the distribution of water ices near the lunar poles as determined by Li et al. (2018) using data obtained by NASA's Moon Mineralogy Mapper (M3) instrument aboard the already quoted Chandrayaan-1 Indian spacecraft. The gray scale corresponds to surface temperature, darker shades representing colder areas and lighter shades indicating warmer zones. The ice

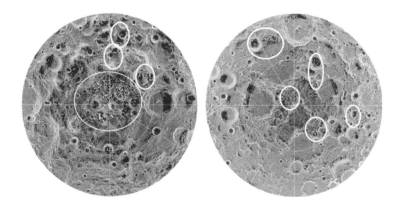

FIGURE 6.1 The distribution of surface ice at the Moon's South Pole (left) and North Pole (right). White circles and ovals indicate the regions with high water ice content. Credits: NASA. Figure adapted from the original in: https://www.jpl.nasa.gov/news/news.php?feature=7218& utm_source=iContact&utm_medium=email&utm_campaign= nasajpl&utm_content=daily20180820.

is concentrated at the darkest and coldest locations inside shadowed craters. These results are the first definitive evidence of lunar water ice.

The instrument collected the reflective properties expected from ice and was able to measure the distinctive way its molecules absorb infrared light, so it could differentiate between liquid water or vapor and solid ice. With enough ice sitting within a few millimeters from the top surface, water would be accessible as a resource for future human expeditions to the Moon.

All major space agencies and several associations are active in promoting lunar colonization for scientific and commercial purposes. Some examples are provided in the following.

The International Lunar Exploration Working Group (ILEWG, http://sci.esa.int/ilewg/) is a public forum sponsored by the world's space agencies to support international cooperation toward a world strategy for the exploration and utilization of the Moon.

The Moon Village Association (MVA, https://moonvillageassociation.org/) is a non-governmental organization based in Vienna. MVA fosters cooperation for existing or planned global Moon exploration programs, be they public or private initiatives.

Again, on the European side, ESA is deeply involved in raising awareness of the public at large and interest of industrial consortia. Among the many documents, the following movie can be quoted: http://m.esa.int/spaceinvideos/Videos/2015/01/ Destination Moon, which contains a historical summary of earlier enterprises in addition to future developments.

NASA established in 2004 a Lunar Exploration Analysis Group (LEAG, https://www.lpi.usra.edu/leag/) to provide analysis of scientific, technical, commercial, and operational issues. LEAG supports lunar exploration objectives and their implications for architecture planning and activity prioritization. Moreover, a "Back to the Moon" program is under way. Perhaps a first step will be a lunar space station known as the Lunar Orbital Platform-Gateway, inhabited by four international crewmembers. The Gateway, a smaller version of the ISS is meant to help humanity extend its footprint out into deep space, and also to enable a variety of interesting scientific and commercial activities on and around the Moon. If everything works out, NASA astronauts could set foot on the Moon before the end of the 2020s. For more details see: https://www.nasa.gov/topics/moon-to-mars/lunar-outpost.

In preparation for a landing on the far side of the Moon, the Chinese National Space Administration (CNSA, see https://www.spacetv.net/cnsa/) placed a communication satellite, named Queqio, in an orbit around the L2 point of the Earth–Moon (E–M) system. The orbital plane of Queqio is perpendicular to the E–M line and the radius described by the satellite around this line is approximately 60,000 km, thus allowing communications to/from Earth with a transmitter/receiver placed on the lunar far side surface. The subsequent mission, Chang'é 4, launched on Dec. 7, 2018 atop a Long March 3B rocket, touched down within the Von Kàrman crater inside the South Pole–Aitken basin on January 3, 2019. During this historical mission, which comprises scientific and technical components provided by Dutch and German institutes, the lander and the rover will study both the surface and subsurface of this region. The rover is named Yutu-2[1], as its predecessor delivered by Chang'é 3 to the near side. The lander carries a biological experiment, namely a small canister containing silkworm eggs and the seeds of several plant species, including potatoes and *Arabidopsis thaliana*, which is a model organism for plant biology and genetics. The aim is to study if and how the plants will support the silkworms with oxygen, and the silkworms will in turn provide the plants with necessary carbon dioxide and nutrients through their waste in the low-gravity lunar environment. At the time of writing, only a few days after the successful landing, news is coming that the biology experiment is indeed working; in other words, there is life today on two bodies of the Solar System, Earth and the Moon. The future months will tell us more.

Furthermore, very low-frequency studies of the Universe will be carried out. The construction of radiotelescopes on the far side of the Moon was vigorously promoted by the late French astronomer

Jean Heidmann, and has been recently advocated by Joseph Silk (2018). The lack of ionosphere (thus opening the last unexplored electromagnetic channel of the longest radio waves) and shield of terrestrial interferences promise great scientific advantages.

China's lunar program will continue with the launch of Chang'é-5, a much advocated sample-return mission. For sure, the Chinese program plans to send humans to the Moon, and perhaps Chinese astronauts will be the first to set foot again on the lunar soil.

The South Korea Aerospace Research Institute (KARI, https://www.kari.re.kr/eng.do) is interested too in lunar human colonies. Their solution is to take advantage of a deep lava crater to protect the inhabited base, as shown in the movie: https://www.youtube.com/watch?v=dYXrUodV9ok.

See Gibney (2018) for a nicely illustrated report of current plans for Moon bases.

In the future, maybe the Moon will act as a refueling station or even as the launch site for missions to Mars and outer Solar System bodies, thanks to the reduced gravity. In a vivid way of speaking, the Moon is about 95% to infinity. Many experiments to directly produce the needed components of space ships on site, with the help of 3D printers, are pursued by National Agencies and private consortia.

6.2 MARS

Mars is the obvious target for both aspects of extra-terrestrial life, alien and our on other worlds. Its diameter of 4,250 km is half-way between Earth's and Moon's. The gravity on the surface is about 63% lower than terrestrial, namely 3.7 m/s^2. The composition of the atmosphere is quite different from ours, being dominated by CO_2, with no more than 2% of argon, 2% of nitrogen and traces of the rest. The atmospheric density, about 100 times thinner than ours, is not sufficient to protect from sterilizing solar wind and UV radiation nor micrometeorites. It is dense enough though to sustain flights of airplanes and balloons. As for Venus, there is no overall magnetic field protecting the surface from solar wind particles (see Figure 6.2).

The surface temperature varies between −140 C and +20 C. In the present epoch, the obliquity of the rotation axis is very similar to that of Earth, so that Mars has four seasons in a year that lasts 637 terrestrial days. The duration of the day is also very similar to ours (1 Martian day = 1 sol = 23 h 37 m). Such an orbit presents a communication problem, because for several weeks, Mars is in the other side of the Sun with respect to Earth, and it cannot transmit nor receive signals directly to/from Earth. An interplanetary communication infrastructure is needed for permanent human colonies.

The presence of water ices on the surface is amply documented by spacecraft images. For instance, the poles of Mars have huge caps composed primarily of water ice below a thin layer of CO_2 ice. The water ice was deposited in successive epochs in layers that contain varying amounts of dust. Thanks to canyons that dissect these deposits, orbiting spacecraft can view the layered internal

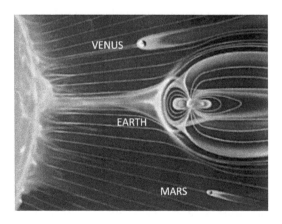

FIGURE 6.2 Earth is protected against the energetic particles of the solar wind by a large-scale dipolar magnetic field. The Moon enters in such screen for five days at each lunation. Venus and Mars do not have such protection (figure not to scale).

structure. In addition to the splendid images of the NASA space-craft and ESAs MarsExpress, the Color and Stereo Surface Imaging System CASSIS aboard the ExoMars 2016 orbiter (for a description of CASSIS see Thomas et al., 2017) started in mid-2018 to provide 3D images of the surface. Underground water ice and even liquid water can be detected by penetrating radar waves. The MARSIS radar carried on board the MarsExpress ESA mission has identified subsurface water deposits (Orosei et al., 2018). Data provide evidence of a lake of liquid water 1.5 km below the ices of the South Pole. The radar is a dual channel low-frequency sounder, operating between 1.3 and 5.5 MHz.

Another specific scientific goal of ExoMars 2016 is the investigation of the Martian atmospheric methane. On Earth, the majority of methane comes from bacterial activity. On Mars atmosphere, the lifetime of methane is very short, around 150 years due to dissociation from solar UV radiation. A continuous variable source near the surface is therefore necessary. The presence of underground bacteria is a fascinating hypothesis. An alternative explanation is the interaction of warm rocks with subsurface liquid water. For a thorough discussion, see Etiope (2018).

A second radar, SHARAD, designed as MARSIS by the University of Rome La Sapienza, is carried onboard NASA's Mars Reconnaissance Orbiter (MRO). Like MARSIS, SHARAD is capable of detecting dielectric discontinuities in the subsurface. See for instance Castaldo et al. (2017) for a global permittivity map of the surface of Mars and its correspondence with geological features.

Three missions plan to penetrate the soil of Mars, two from NASA (InSight, Mars 2020) and one from ESA (ExoMars 2020).

InSight (for Interior Exploration using Seismic Investigations, Geodesy and Heat Transport), landed on November 26, 2018 in the northern portion of flat-lying Elysium Planitia about four degrees north of Mars' equator. InSight will study the deep interior of Mars. The lander's instruments include a seismometer to detect marsquakes and a probe that will monitor the flow of heat from the planet's interior. Although not directly involved in the search of life, nevertheless InSight will provide precious information on the underground conditions where bacteria may have lived.

The Mars 2020 mission includes in its key science goals, the capability to answer questions about the potential for life on the planet. The mission will seek not only signs of habitable conditions in the ancient past, but also signs of past microbial life itself. A drill on a rover will collect core samples of the most promising rocks and soils. Furthermore, Mars 2020 will gather knowledge and demonstrate technologies in view of future human expeditions to Mars.

ESA's ExoMars 2020 shall target a location[2] interpreted to have strong potential for past habitability and for preserving physical and chemical biosignatures, as well as abiotic/prebiotic organics. The mission will deliver a lander with instruments for atmospheric and geophysical investigations and a rover tasked with searching for signs of extinct life using a drill capable of collecting material from outcrops and at depths down to 2 m. This subsurface sampling capability will provide the best chance yet to gain access to chemical biosignatures (Vago et al., 2017).

After the pioneering and at the first sight of disappointing data of the Viking spacecraft in the 1970s, the presence of building blocks of life on Mars is treated by a very large amount of literature. Recent re-examinations of the Viking data are not as negative with respect to the capability to support life as in the first interpretations.

Among the findings obtained by subsequent missions, we quote those gathered by NASA's Curiosity rover by drilling a 5 cm deep hole in a Martian rock located inside the Gale crater. Analysis of the rock reveals that 3.5 Gy ago, the lake that once filled Gale Crater contained complex organic molecules. Those molecules are preserved still today, possibly thanks to sulfur that may have protected the organics even when the rocks were exposed to radiation and bleaching perchlorates.

Summing up all the evidence collected by ground-based telescopes and space missions we can conclude that all fundamental building blocks of life were and still are present on Mars. However, there are also elements adverse to the presence of life, at least at the present epoch:

- The first adverse element is the already quoted absence of an ordered magnetic field shielding from the solar wind. Moreover, the thin atmosphere does not protect from sterilizing UV radiation nor infalling small meteorites. The spacecraft MAVEN (Mars Atmosphere and Volatile Evolution mission, https://mars.nasa.gov/maven/; http://lasp.colorado.edu/home/maven/) has among its main scientific goals to ascertain the fate of the original atmosphere and magnetic field of Mars. Jakosky et al. (2018a) determine that the integrated loss of gases, based on MAVEN-measured rate and adjusted according to expected solar evolution, are equivalent to at least 23 m of a global layer of H_2O. Combined with the lack of surface or subsurface reservoirs for CO_2 that could hold remnants of an early, thick atmosphere, these results suggest that loss of gas to space has been the dominant process responsible for changing the climate of Mars from an early, warmer environment to the cold, dry one that we see today.

- The lack of a massive moon does not stabilize the obliquity of the rotation axis of Mars. In a few hundred thousand years, today's equatorial regions will be almost polar. No stable conditions for life evolution exist.

- From time to time, the entire planet is engulfed by massive dust storms, as seen in several occasions since Mars is observed from ground telescopes. The most recent episode occurred in April 2018 and lasted until the next August, causing some problems for Curiosity. Those month-long episodes of overcast sky by dust speckles may have endangered the Martian life and will constitute a risk for human outposts.

In conclusion, the present data do not provide unambiguous evidence for past nor present life on Mars. Future missions will tell more.

Regarding human life on the planet, colonization of Mars is advocated by illustrious personages, such as the late Stephen Hawking and the founder of the Tesla and SpaceX companies Elon Musk, for quite different motivations, but surely tending to the same objective. As for the Moon, national space agencies and private groups are preparing structures and people for the long journey and inhospitable environment. For instance, the Austrian-led Forum OeWF (http://oewf.org/en/portfolio/amadee-18) conduct field training expeditions such as AMADEE in Oman in 2018 (see Figure 6.3, taken from https://oewf.org/en/portfolio/amadee-18/).

As examples of Mars habitats, we can quote the MIT concept that won the top prize for architecture in 2017's international Mars City Design competition, sponsored by both NASA and ESA (https://www.marscitydesign.com/about).

Another interesting concept is the "igloo" of the Swiss École Polytechnique Féderale de Lausanne (EPFL, https://actu.epfl.

FIGURE 6.3   Two members of the AMADEE Expedition walk in in the desert. Image credit OeWF Florian Voggeneder.

ch/news/scientists-sketch-out-the-foundations-of-a-colon-2/). The igloo has been designed for a location near the North Polar regions of Mars. The Martian poles contain perchlorate, a substance used as a rocket fuel but also by some microbes as a source of energy. According to the concept outlined by researchers at EPFL, the base would host humans on the surface for a 288-day stretch. During that time, the North Pole is bathed in sunlight continuously. The outpost will be covered by a dome made of polyethylene fiber shrouded in a three-meter layer of ice like an igloo. Under the dome, there are living spaces and airlocks leading to the exterior. The dome protects explorers against radiation and micrometeorite strikes and keeps internal atmospheric pressure constant as well.

Concluding this chapter, humanity seems to be on the verge of large-scale colonization of outer space, both by nations and by private, powerful and often transnational conglomerates motivated essentially by favorable economic perspectives. However exciting the projects, the difficulties are tremendous. See for instance Jakosky et al. (2018b) for a realistic appreciation of the possibilities of terraforming Mars with available technologies. While returning human crews to the Moon appears feasible in the next decade, the date for a human landing on the red planet is possibly two or three decades ahead.

International cooperation seems mandatory. Rules should be agreed, implemented and enforced by a multinational authority if peaceful colonization is sought. The scenario of frequent commercial flights might offer new possibilities for science. In past centuries, scientists were taken as host passengers on board vessels going to unexplored lands, again for commercial reasons. That situation might happen in the future. No selection committees nor endless paper work to have a research project approved, just a free ride for serendipitous discoveries.

NOTES

1. A gentle Chinese legend interprets the pattern of bright-and-dark features visible by eye on the lunar near side as the rabbit Yutu bringing a cup of a magic potion to the lunar princess Chang'é.
2. Not definitely chosen at the time of writing.

CHAPTER 7

Comets, Asteroids, Meteorites, Death and Life

COMETS HAVE BEEN OBSERVED for millennia and surely not only for scientific motivations. Comets indeed can be fascinating visions on the sky. They were, and for many still are, an enigma, a transient mysterious phenomenon perturbing the orderly heavens. The apparitions were associated with important events such as the birth or death of a prince, battles, famines and pestilences. Therefore, the appearances of great comets were recorded in several historical documents across the entire world, from Asia to Europe to America. In many cases, the astrologers described the cometary aspects, their color and their paths among the constellations, all elements of great importance later on for the foundation of modern cometary science. A paradigmatic case is Halley's: all passages of this bright comet were observed and recorded since at least year 240 BC and perhaps even earlier. From the more recent passages, in 1705 Edmond Halley was able to calculate that the orbit had a periodic characteristic with a period of about 76 years and predicted its return in 1759. When the prediction came true (Halley was already dead at the time), the comet was named Halley's, 1P, the first comet recognized to be periodic.

For a review of cometary historical facts, development of scientific knowledge and contemporary vision see Barbieri and Bertini (2017) and the references therein.

Let us briefly consider the association of cosmic bodies to the widespread death of living species on Earth. Several great extinctions of life occurred in the past eras, the extinction of the dinosaurs being possibly the best known, but surely not the most dramatic. The worst of all happened some 250 My ago (more or less where the Sun is today in its galactocentric orbit), at the boundary between the Permian and Triassic periods. This dramatic extinction most likely was due to a massive eruption of molten basalt in Siberia, not to the infall of a cosmic body (see for instance https://physicstoday.scitation.org/do/10.1063/PT.6.1.20180905a/full/). In other cases, episodes of mass extinctions can be associated with great impacts of cosmic origin. According to several authors, the occurrence of such large impacts seems to show a periodicity around 30 My. If true, what is the reason for the periodicity? We have already mentioned the pendulum motion of the Sun above and below the galactic plane with such a period. According to authors defending the periodicity of impacts, the larger density of matter encountered when crossing the galactic plane favored larger impacts. There is another theory about the supposed periodicity, which postulates that the Sun is not a single star. There is a very faint companion, named Nemesis, in a very large elliptical orbit. When Nemesis approaches the Oort cloud of comets, its perturbations favor the infall of comets to the inner Solar System. Actually, Meier et al. (2017) do not find any periodicity in the best-recorded impact episodes of the latest 250 My.

After the extinction of the dinosaurs and the great reptiles, great mammalians and finally humans appeared. Impacts continued too, albeit at a much-reduced rate, and happen still today. The Tunguska event of 1908 is one of the most widely known.

The danger caused by infalling cosmic bodies cannot be under-evaluated, as demonstrated for instance by Yeomans (2013). Several strategies have been conceived to avoid impacts, or at least mitigate their effects. Whatever the solution, time of discovery is a most critical parameter; the sooner the dangerous body is discovered the better. Therefore, worldwide efforts are vigorously performed, by radio and optical means, to discover bodies with orbits intersecting that of Earth and to estimate the hazard of impacts. A summary of the efforts performed by NASA in the last 20 years has been published in: https://www.jpl.nasa.gov/news/news.php?feature=7194&utm_source=iContact&utm_medium=email&utm_campaign=NASAJPL&utm_content=daily20180723-3.

However, recent episodes reveal that bodies with sizes under 100 m or so can arrive undetected until very close approaches, too late for any measure. The surveillance network needs to be improved.

Enough with disasters, indeed great explosions of life occurred too! Two galactocentric rotations ago (600–500 My ago, at the transition between Ediacaran and Cambrian), there was an extraordinary explosion of life on Earth. Quite possibly, extremely well preserved Dickinsonia fossils found in northwest Russia near the White Sea represent the oldest known animal. Some 1.5 m long, these creatures appear to have had organic tissue containing molecules of cholesterol, a marker found only in animals (Bobroskiy et al., 2018). The following period, the Cambrian, witnessed a truly impressive step in the evolution of living beings, with the appearance of the first structured animals, with skeletons, legs, fins and other apparatuses useful for mobility, offense and defense.

Among the causes of such enrichment of life is certainly the increased content of oxygen in the atmosphere. Several authors maintain that the importance of oxygen is far from linear, with steps at 1% and 10%. Why did oxygen's content increase? There are several reasons: plate tectonics of course, but infalling cosmic bodies had their influence too.

We have already seen that a living organism needs:

- Four basic chemical elements: H, C, N, O
- Twenty amino acids (molecules made of $-NH_2$ + $-COOH$) with left-handed symmetry (a cosmic origin for such asymmetry is widely discussed, as already mentioned)
- Liquid water, plus a source of energy for metabolism

Could a fraction of these basic ingredients come from the outside, born in interstellar clouds and carried to Earth by comets, asteroids and meteorites? Some lines of evidence are given in the following.

The interstellar clouds of dust and gases are extra-terrestrial laboratories, where chemical, physical and maybe biological reactions occur, which are almost impossible to reproduce on Earth and which touch the very heart of our review, namely the origin and development of life. Notice that the birth of complex molecules and prebiotic components in interstellar clouds has at its disposal much more time than the 4.6 Gy provided by our Earth.

A large number of atoms and molecules (about 200 at the present time, but the inventory is far from exhausted), from atomic to molecular hydrogen to molecules made of 13 atoms, have been discovered inside those very cold and rarefied clouds (say 25–40 K, a dozen of molecules per cubic centimeter). Table 7.1 gives an incomplete list of those atoms and molecules.

Molecules containing silicon have been inserted in the Table, with the caveat that their number is only a few percent of the total, in order to show again that carbon is much more abundant in the Universe than silicon, a preferred element by science-fiction writers for alternative forms of life. Table 7.1 is continuously updated. Some examples of new entries are given in the following:

- Ohishi et al. (2017) report the detection of Methylamine CH_3NH_2 (a possible precursor to glycine) in celestial sources using the Japanese Nobeyama 15-m radiotelescope.

- Formamide (NH_2CHO) has been observed in interstellar nebulae using IRAM/NOEMA (see later for a description of the instrument): this molecule is important as a possible starting point for the formation of metabolic and genetic macromolecules. The discovery is part of a large project called SOLIS (Seeds of Life in Space, see Ceccarelli et al., 2017).

- Using ALMA, the first detection in a protoplanetary disk of formic acid (HCOOH, the simplest organic acid) has been carried out by Favre et al. (2018).

- The already quoted paper by McGuire et al. (2018) reports, in addition to HDO and CS the detection of the complex organic molecule glycolaldehyde ($HC(O)CH_2OH$).

Comets and asteroids, which are primitive, essentially unevolved bodies, carry precious information on those great interstellar clouds of dust and gases. They transport to the planets, in particular to Earth, molecules of the four basic ingredients, H, C, N and O, as exemplified by Figure 7.1.

Comet 46P Wirtanen is a member of the Jupiter family and orbits around the Sun every 5.4 years. It was the original target of the Rosetta mission, but a problem with the Ariane V launcher imposed a delay of one year, and a change of comet and asteroids. The Rosetta mission is described in detail later on. Many other elements, like phosphorus, sulfur and iron, rare gases such as xenon, organic compounds and even more complex molecules such as amino acids are found in the gases of the comae and tails of comets.

Very recently, observations proved a long-suspected hypothesis, namely that some comets and/or asteroids can come inside our Solar System from farther out. Object A/2017 U1, discovered in 2017, made a very rapid passage through the inner Solar System. It made its closest approach to the Sun on September 9, 2017 moving with a velocity larger than the escape velocity at that distance. Therefore, the object most likely is of interstellar origin,

TABLE 7.1 A List of Some Interstellar and Cometary Compounds

Category									
Hydrogen compounds	H	H_2	H_3^+	H_5^+					
Hydrogen and carbon compounds	C_2	C_3	C_5	CH	C_2H_4	H_2C_4	$H2C_6$	C_8H	
Hydrogen, carbon and oxygen compounds	OH	CO	CO^+	H_2O	HCO	C_2O	H_2CO	CH_3COOH	CH_3O_2CO
Hydrogen, carbon and nitrogen compounds	NH	CN	NH_2	HCN	NH_3	H_2CN	CH_2CN	CH_3CN	$HC_{11}N$
Sulfur containing compounds	SH	CS	SO	NS	H_2S	SO_2	H_2CS	CH_3SH	C_5S
Silicon containing compounds	SiH	SiC	SiO	SiS	$HSiC_2$	SiH_4	$HSiC_2$	SiC_3	SiC_4

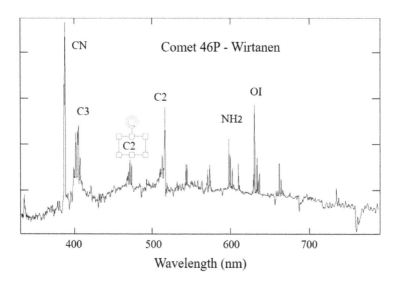

FIGURE 7.1   A low-resolution spectrum of comet Wirtanen taken in December 2018 near its perihelion at the Asiago Observatory (Courtesy P. Ochner, Unipd - INAF OAPd).

and thus it was renamed 1I/2017 U1, where 1I means first interstellar. Its nickname is Oumuamua, namely a messenger from far away, in the Hawaiian language. By combining data from the Hubble Space Telescope, the Canada–France–Hawaii Telescope, ESO's Very Large Telescope and the Gemini South telescope, it was found that Oumuamua was moving faster than predicted. The most likely explanation is that Oumuamua was outgassing material from its surface, a behavior typical of comets (Micheli et al., 2018). Where did Oumuamua come from? Gaia data suggest four plausible candidates (Bailer-Jones et al. 2018), all nearby dwarf stars, but with a high degree of uncertainty.

7.1 THE IMPORTANCE OF INFRARED AND RADIOTELESCOPES

Studies of interstellar clouds, comets, asteroids, namely of all cold objects, are greatly helped by infrared and radio telescopes. The emitting or absorbing matter indeed is very cold, say 15–100 K, and the peak of Planck's Black Body curve is in the far infra-red (IR) or millimeter radio region. An example is the famous Barnard's nebula (Figure 7.2), which is absolutely impenetrable to optical radiation, but it is full of molecules that can be detected by far-infrared and radiotelescopes.

Many ground telescopes are equipped with excellent IR detectors able to cover the spectral band from, say, one to five micrometers. Their main limitations are due to the varying transparency and high thermal background of the terrestrial atmosphere. IR space telescopes do not have the terrestrial atmosphere limitation. Furthermore, the entire telescope structure can be encased in a cold vessel, e.g. filled by liquid helium, thus eliminating the noise due to the telescope structure itself. The choice of an orbit distant from Earth (e.g. in the Sun–Earth L2 Lagrangian point) and efficient shielding reduce to a minimum the heat received from the Sun and Earth. Their main limitation is the finite operational time. When the cooling agent finishes, the telescope terminates its operational "cold" time. In some cases, it can be put in a "warm" mode for further observations of less sensitivity and spectral band coverage.

FIGURE 7.2 The Barnard's nebula, impenetrable to optical telescopes.

Therefore, IR space telescopes are major players in studying cold matter. We wish to quote here the now defunct European ISO and Herschel telescopes. ISO (http://sci.esa.int/iso/) operated between 1995 and 1998 at wavelengths from 2.5 to 240 microns and provided interesting data on the amount of water in the Universe. Herschel (http://sci.esa.int/herschel/) flew the largest single mirror ever built for a space telescope. The 3.5 m diameter mirror collected long-wavelength radiation covering a spectral range from the far infrared to sub-millimeter. Among the immense amount of literature generated by Herschel, we quote a recent reanalysis (Rice et al., 2018) of data toward two protostellar sources. Such paper sheds further light on the debate whether the original carriers of Earth's nitrogen (a key element for life and primary component of the atmosphere) were molecular ices or refractory dust. The analysis reveals that HCN is the organic molecule that contains by far the most nitrogen and that refractory dust seems to be the bulk provider of nitrogen to comets.

Another important IR telescope is NASA's Spitzer, with a mirror of diameter 0.85 m (http://www.spitzer.caltech.edu/). Supposed to last five years in its Earth trailing orbit, after 15 years the telescope is still providing excellent data in its "warm" operation mode. The future NASA 6.5 m folded mirror James Webb Telescope, tailored for near IR, will undoubtedly produce many excellent results.

Starting around 1960, radio astronomy flourished all over the world, and in the following decades provided an impressive amount of molecular data, because the radio spectrum is very rich with spectral lines and bands due to a great variety of molecules. Those lines and bands are usually due to rotational transitions, simpler to interpret than vibrational or electronic transitions typical of the visual and UV bands; spin temperatures and isotopic ratios can be accurately determined; the measurement of radial velocity is extremely precise. Furthermore, if the object comes sufficiently close to Earth, radars can give additional exceptionally interesting data, providing at the same time position and velocity of the body, its shape and rotational state, even some characteristics of its surface from the features of the return echo.

Among the several radiotelescopes engaged in such studies, we have already mentioned Nobeyama in Japan (https://www.nro.nao.ac.jp/en/) and IRAM (Institut de Radioastronomie Millimétrique, ww.iram-institute.org/). Founded in 1979, IRAM today maintains two observatories: the glorious 30 m telescope located on Pico Veleta near Granada, Spain, and the NOEMA interferometer (currently an array of nine 15 m telescopes) in the French Alps, the most powerful millimeter radiotelescope in the Northern Hemisphere.

The multination ALMA complex (see http://www.eso.org/sci/facilities/alma.html) deserves a second special mention being the largest such facility in the world. Located in Chajnantor (Northern part of the Atacama Desert in Chile), at an elevation above 5,000 m, it enjoys a very low atmospheric content of water vapor. In addition to the large collecting surface ensured by the many aerials, a most important virtue is its capability to phase all signals from the antennae in interferometric mode, achieving a very fine angular resolution in the sky and exquisite sensitivity. Among its main scientific goals is the "In Search of Our Cosmic Origins" project.

Another example is the Five hundred meter Aperture Spherical Telescope (FAST, http://fast.bao.ac.cn/en/) radiotelescope in China. FAST is the largest single dish radiotelescope in the world. Two scientific themes of FAST are connected to the present review:

- Detection of interstellar molecules. The receiver bands of FAST are designed to cover OH, CH_3OH and 12 other molecular lines. FAST's high sensitivity will enable the search for long-chain carbon molecules in our Galaxy, as well as extragalactic OH and CH_3OH molecules

- Detecting interstellar communication signals. If an alien civilization located at a distance of about 28 light-years transmits 1 GWatt of omnidirectional radio power, FAST could detect such transmission. Notice that there are some 1,500 stars inside that volume.

7.2 THE IMPORTANCE OF METEORITES

Nature itself opens for us a second most important channel of information on the Universe, namely matter in addition to electromagnetic waves. Probably the most important collection of meteorites in the world is the one kept at the Vatican Observatory in Castelgandolfo near Rome (Italy).

Dust grains continuously bombard Earth. Small speckles of dust burn in the mesosphere at about 90 km altitude. Occasionally, the size and trajectory of the body allow it to fall to ground and be collected as a meteorite. From their isotopic composition, it has been concluded that some meteorites originated by impacts on the Moon, Mars, the asteroid Vesta and after long journeys in the Solar System fell on Earth, often to be recuperated in Antarctica or in sandy deserts.

The sooner a meteorite is recovered the better, because there is little or no time for contamination by bacteria or humidity in the soil. Among those meteorites, Murchison's, found in Australia in 1969, deserves a special mention. The meteorite, made of many broken pieces, is a rocky meteorite packed with organic molecules. Another founder of astrobiology, the NASA biologist Cyril Ponnamperuma, discovered in it guanine, adenine, cytosine, uracil and thymine. The meteorite carry information about the young Sun, whose past is recorded in blue crystals known as hibonite $(Ca,Ce)(Al,Ti,Mg)_{12}O_{19}$ made of one of the first minerals to form in the Solar System. These specks contain traces of chemical activity from the early periods before any of the planets formed. Using a powerful laser beam, Koop et al. (2018) were able to release neon and helium trapped inside the hibonite crystals. The concentration and ratio of isotopes of these noble gases showed that an energetic young sun irradiated the hibonite crystals present in the cloud of gas and dust around the still-growing Sun. When the

Sun's high-energy particles struck the blue crystals, they split calcium and aluminum atoms to make isotopes of neon and helium. An ancient meteorite made of andesite, possibly the oldest igneous variety ever cataloged, seems to contradict a paradigmatic idea about planets habitability, according to a NASA press release of August 10, 2018 (http://spaceref.com/meteorites-1/meteorite-research-could-reshape-understanding-of-how-planets-form.html). This research goes against the idea that all silica-rich rocks contain water, and therefore the keys to habitability. Since the silica-rich meteorite in the study likely formed without the presence of liquid water, searching for bodies in the universe that have silica-rich rocks as a way to define habitability may no longer work.

7.3 THE IMPORTANCE OF SPACE MISSIONS

Spacecraft orbiting close to celestial bodies and even landing on them, offer crucial advantages with respect to terrestrial telescopes or telescopes in low Earth orbit like the HST. Proximity to the body is obviously a great advantage, as are the absence of terrestrial atmosphere with its limitations in transparency and turbulence and the absence of the night/day cycle. Equally important is the possibility to utilize both channels of information, the electromagnetic one plus the material one, at the same time.

Space missions are crucially important for comets, because of the very complex interactions between the radiation, magnetic field and particles coming from the Sun and the cometary nucleus. Such interactions happen very quickly in a small volume around a nucleus that generally is no more than few km across, and essentially below the spatial and temporal resolution of ground telescopes, as exemplified in Figure 7.3. The figure shows two images of comet 67P C-G taken on the same day in January 2016, one by the Schmidt telescope in Asiago–Cima Ekar and one by the Osiris camera on board ESA Rosetta inside the coma of the comet.

Therefore, spatial and temporal resolutions explain why space missions are of such paramount importance for comets. In

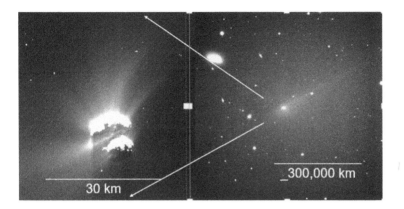

FIGURE 7.3 Two images of comet C-G taken the same day in January 2016. On the right is the image from the Schmidt telescope in Asiago–Cima Ekar Observatory (courtesy P. Ochner, Unipd - INAF OAPd), on the left the image taken from the Osiris camera on board ESA's Rosetta mission. Image credit ESA/Rosetta/MPS for OSIRIS Team MPS/UPD/LAM/IAA/SSO/INTA/UPM/DASP/IDA).

particular, a long duration mission, capable of remaining inside the cometary coma for several months, both before, during and after the perihelion phases. Those were the unique characteristics of the European mission ROSETTA, which will be described in the next chapter.

CHAPTER **8**

The Cometary Mission Rosetta

T HE EUROPEAN COMETARY MISSION Rosetta derived its name
from the famous stone, which is on display in the British
Museum in London. The stone has been a fundamental landmark
in deciphering the ancient Egyptian hieroglyphic scripture, as it
was the Philae obelisk, which gave the name to the landing module of Rosetta. See Barbieri, 2017 and references therein for more
details on the stone. As the stone was decisive for deciphering the
hieroglyphic scripture, so the mission represents a fundamental
landmark for cometary science.

The present chapter will expound some of the results obtained
by Rosetta pertaining to the theme of life, concentrating on two
instruments aboard the spacecraft, namely the imaging camera
Osiris and the mass spectrometer ROSINA.

Rosetta left the launching pad in Kourou aboard an Ariane
5 rocket on March 2, 2004. After three flybys of the Earth, of
Mars and two asteroids (Steins and Lutetia), the spacecraft finally
reached comet 67P C-G on August 6, 2014, just inside Jupiter's
orbit. For the first time in space exploration, a human machine

could spend almost two years around a comet, from outside the orbit of Mars to the perihelion to again outside of Mars orbit. The primary scientific target of the mission, comet 67P C-G, was discovered by two Soviet astronomers, K. Churyumov and S. Gerasimenko in 1969. The comet was captured by Jupiter ten years before its discovery, and since then it has been a member of the Jupiter family of comets (JFC) and revolves around the Sun with a period of approximately 6.5 years.

As already said, the original target of the mission was comet 46P Wirtanen, another member of the JFC. The two comets have different characteristics, 67P being much gas poorer than 46P, and approximately three times larger and nine times more massive. Such differences required some last minute reworking of the spacecraft, in particular a reinforcement of the legs of the landing module Philae due to the stronger gravity.

After the very successful flybys of Mars and the two asteroids, during which many new scientific results were obtained, Rosetta finally reached the comet. When the distance to the comet went below 3,000 km, a very complex body was seen in great detail by the Osiris cameras. The comet is composed of two sub-units (nicknamed head and body) joined by a sort of neck (Figure 8.1). It is most likely that the two parts were originally independent bodies which merged together after a gentle, low-velocity impact.

The complex nucleus rotates with a period of approximately 12.4 hours, with continuous activity of dust and gas jets seen even at great distances from the Sun both before and after the perihelion. During the long mission, the comet displayed several spectacular and unexpected features, like boulders flying away and slides of entire blocks from steep walls. See Barbieri and Bertini (2017) for further details and images. Here, we concentrate in particular on the presence of water ice patches in various places of the surface and jets with a mixture of dust grains and water vapor molecules.

Water ice in comets has several peculiar properties. For instance, the water molecules act as clathrates hydrates, namely as

FIGURE 8.1   Image of comet 67P C-G obtained by the Osiris camera onboard Rosetta. Image credit ESA/Rosetta/MPS for OSIRIS Team MPS/UPD/LAM/IAA/SSO/INTA/UPM/DASP/IDA.

cages encapsulating and protecting other species, which are then released when temperature and pressure conditions are appropriate for opening the cage. The jets seen coming out of 67P even long after the outbound crossing of the orbit of Mars (therefore at a large distance from the Sun) indicate other properties of water and the effectiveness of a still ill-understood internal source of energy. Let us consider in detail the event of 3 July 2016 (Agarwal et al., 2017, see Figure 8.2), when several instruments of Rosetta detected an outburst event at a heliocentric distance of 3.32 AU from the Sun, outbound from the perihelion. The activity coincided with the local sunrise and continued for several tens of minutes. The outburst left a 10-meter-sized shallow icy patch on the surface.[1]

The ejected material comprised refractory grains of several hundred microns in size, and sub-micron-sized water ice grains. The measured high production of dusty material in a short time is incompatible with the free sublimation of crystalline water ice under solar illumination as the only release and acceleration process. Additional energy stored near the surface must have increased the gas density. Two further possible mechanisms are

FIGURE 8.2   Wide angle camera image of the outburst plume obtained on July 03, 2016. Adapted from fig.2 of Agarwal et al. (2017). Image credit ESA/ Rosetta/MPS for OSIRIS Team MPS/UPD/LAM/IAA/SSO/ INTA/UPM/DASP/IDA.

a pressurized subsurface gas reservoir or the crystallization of amorphous water ice. The second alternative seems more attractive. Amorphous ice may have been present behind a thin surficial wall and transformed into crystalline ice when, upon local sunrise, either a part of the wall collapsed or a newly formed thermal crack exposed it to solar irradiation. It is known that below T = 160 degrees Kelvin (K), ice has an amorphous structure, while at temperatures around T = 200 K and low pressure, ice assumes a metastable cubic structure before transforming into the stable hexagonal ice. The transformations from amorphous to hexagonal to cubic are exothermic. Therefore, the temperature increase caused by increasing insolation may have induced such phase transitions. The released energy may have been sufficient to raise the sublimation rate to a level consistent with the observed dust production rate and velocities.

Regarding the mass spectrometer ROSINA, Table 8.1 provides a partial inventory of detected cometary molecules in the gases coming out of the surface.

TABLE 8.1   Partial Inventory of Gases Found by ROSINA

Name	Composition
Water	H_2O
Carbon Monoxide	CO
Carbon Dioxide	CO_2
Ammonia and	NH_3
protonated ammonia	NH_4^+
Methane	CH_4
Methanol	CH_3OH
Formaldehyde	CH_2O
Hydrogen Sulfide	H_2S
Hydrogen Cyanide	HCN
Sulfur Dioxide	SO_2
Carbon Disulfide	CS_2
Molecular Oxygen	O_2
Molecular Nitrogen	N_2
Argon	Ar

The detection of molecular oxygen confirms the very low temperature in the region where the comet originally formed. With its high mass resolution and sensitivity, ROSINA was able not only to detect deuterated water HDO, but also doubly deuterated water, D_2O, water with different O isotopes and deuterated hydrogen sulfide HDS (Altwegg et al., 2017). The ratios for [HDO]/[H_2O], [D_2O]/[HDO] and [HDS]/[H_2S] are $(1.05 \pm 0.14) \times 10^{-3}$, $(1.80 \pm 0.9) \times 10^{-2}$ and $(1.2 \pm 0.3) \times 10^{-3}$, respectively. It is worth remembering that the D/H ratio is equal to half the HDO/H_2O ratio because water contains two atoms of hydrogen. These results yield a very high ratio of 17 for [D_2O]/[HDO] relative to [HDO]/[H_2O]. Statistically one would expect just 1/4. Such a high value can be explained by cometary water coming unprocessed from the very cold presolar cloud, where water is formed on grains, leading to high deuterium fractionation. The percentage of deuterated water in comet 67P (namely the ratio HDO/H_2O) is 2–5 times higher than in the Earth's oceans (the already recalled VSMOW value of D/H = 1.5576×10^{-4}). Moreover, this percentage is 20 to 50 times the value in the Standard Big Bang model, in the presolar nebula and large planets. These results are a further indication of fractionation aided by low temperatures. For a more detailed account and other references, see Spiga et al. (2019). The high [HDS]/[H_2S] ratio is compatible with upper limits determined in low-mass star-forming regions and also points to a direct correlation of cometary H_2S with presolar grain surface chemistry.

In addition to the above results, ROSINA detected glycine, the simplest of the 20 amino acids fundamental for living beings, present in almost all proteins, together with methylamine, ethylamine, phosphorus and a multitude of organic molecules. The detection of glycine by ROSINA was free from possible contamination in the samples collected by the Stardust mission in its flyby of comet Wild 2.

The connection between comets and Astrobiology after Rosetta is discussed by Altwegg et al. (2016) and by Cottin et al. (2017b). As already stated, comets are reservoirs of a large amount of material considered necessary for the origin of life on Earth. While the measurement of the D/H ratio in the water of comet 67P established that similar comets are probably not a significant source of water on Earth, the nature and amount of the volatile organic content of comet 67P, the detection of glycine and phosphorous atoms have demonstrated the presence of so-called "prebiotic" ingredients.

Combining ROSINA and ALMA data, the presence of the organohalogen Freon-40 (CH3Cl, also known as methyl chloride and chloromethane) in the gas around comet 67P was ascertained. Furthermore, ALMA detected Freon around the solar-type double star IRAS 16293-2422. Organohalogens are likely constituents of the so-called "primordial soup," both on the young Earth and on rocky exoplanets. These discoveries strengthen the links between the pre-biological chemistry of protostars and surrounding nebulae and our own Solar System, a link that is becoming one of the most promising fields of investigation.

To conclude this chapter, we recall the comprehensive review of pre-Rosetta isotopic ratios in comets given by Bockelée-Morvan et al. (2015). Rosetta has provided additional information and confirmed that comets contributed in a substantial way to the presence of prebiotic elements and rare gases in the primitive terrestrial atmosphere. Such elements were already present in the dust grains and gases of a presolar nebula and subject to a variety of processes, due not only to interaction with UV light from hot stars and high-energy cosmic rays but also within the atmospheres of cool low-mass stars, where most of the known habitable exoplanets are found. Therefore, let us leave our Solar System, moving beyond the orbit of Pluto, the Kuiper Belt and the Oort cloud of distant comets, where the gravitational force of the Sun is almost balanced by that of the nearer stars. The first stars are encountered beyond one million AUs. The next chapter will discuss the present knowledge about their planets.

NOTE

1. During this orbital passage, the comet lost a shell with a thickness equivalent to about 70 cm, with appreciable differences from place to place according to the duration of the insolation.

CHAPTER 9

Planets of Nearby Stars

A FTER THE RADIO DETECTION of a planet around a pulsar in 1992, the first optical exoplanet orbiting the solar-type star 51 Pegasi was discovered in 1995 by the radial velocity variation of its host star (Mayor and Queloz, 1995). Since then, the number of exoplanets has grown continuously. Today, their census surpasses 4,000, thanks in particular to the NASA Kepler satellite. Most of those exoplanets are gaseous giants similar to Jupiter or even larger, but a sizeable fraction have an Earth-size diameter. A continuously updated list of exoplanets and their properties can be found in the Extrasolar Planets Encyclopaedia (http://exoplanet. eu/catalog/) and in the Mikulski Archive for Space Telescopes (MAST, http://archive.stsci.edu/archive_news/2018/08-Aug/ index.html#article1) maintained at the Hubble Space Telescope Science Institute.

The astrometric satellite Gaia is already contributing information about the habitability of hundreds of exoplanets (see for instance Johns et al., 2018), but the importance of Gaia will increase with time. The reason is that the accuracy of measurements of the proper motions of the host stars and of the tiny wobble induced by the planet will increase with time. Therefore, the determination of the mass of the planet, perhaps in conjunction with Hipparcos data obtained more than 25 years ago, will become better and better. Such an astrometric method has been applied for the first time to the measurement of the hot-Jupiter accompanying Beta Pictoris, see Snellen and Brown (2018).

The nearest exoplanet known today is in the southern Milky Way, inside the triple star system of Alpha Centauri, a mere 4 light-years (l-y) away. Proxima, the nearest of these three stars, has a planet, Proxima-b, with some similarities to our Earth, although it orbits much closer to its faint cool star (spectral-type M) with an orbital period of 11 days. ALMA has detected the presence of a ring of cold dust (≈ 40 K, total mass $\approx 1/100$ Earth masses) around Proxima, in a region between one and four AUs. The data also hint at the presence of an even cooler outer dust belt. These structures are similar to the Main Asteroid Belt and Kuiper Belt in our Solar System, albeit on a much smaller spatial scale, and are expected to be made from particles of rock and ice that failed to form planets. There are other exoplanets with dust rings and presumably comets and asteroids around them. For instance, the star Eta Corvi (spectral type F2-V, distance 60 l-y), has a similar ring of matter around it, according to infrared observations by the NASA telescope Spitzer. The presumed age of Eta Corvi is 1 Gy, more or less the same age of Earth during the Late Heavy Bombardment.

The greater part of planets in the habitable zone are found around low-mass cool M-type stars. In addition to Proxima-b, two more cases are worth mentioning:

- A super-Earth orbiting Barnard's star (Ribas et al., 2018). Barnard's star is also very near to us, only 6 l-y away. The star has also the highest known proper motion (about 10 arcsec/y) with respect to the background of distant "fixed" stars.

- Trappist-1, a system of seven almost co-planar planets orbiting an M-star 39 l-y away (Gillon et al., 2017). This amazing system will be treated again in the next chapter.

9.1 HOW TO DISCOVER AN EXOPLANET

A word of caution: the way the media report the discovery of an exoplanet can lead many to believe that the telescopes obtain real images of it. Except in very few cases, due to adaptive optics coronagraphic systems[1] like SPHERE at the ESO VLT (https://www.eso.org/sci/facilities/paranal/instruments/sphere.html), this is not true! The presence and characteristics of such exoplanets are inferred via two main methods[2]:

1. Variation of the radial velocity of the star due to its orbit around the barycenter of the system star-planet(s). Such variations are very small, not more than a few m/s, and only sophisticated equipment at large telescopes permits their measurement. The 3.5 m Telescopio Nazionale Galileo (TNG, http://www.tng.iac.es/) in the Canary Islands is one of the best telescopes for such measurements in the Northern hemisphere. Figure 9.1 shows the TNG radial velocity variation due to the already mentioned first discovered planet around 51 Pegasi.

The case of 51 Pegasi is one of the easiest because of the high amplitude of the velocity variation. To reach a substantial number of exoplanets the sensitivity of the instrument must reach a few centimeters per second. Therefore, SARG was substituted with a newer instrument, named HARPS-N@TNG (for a complete description see Cosentino et al., 2012). HARPS-N capitalizes on the technical and scientific achievements of the original HARPS (High Accuracy Radial Velocity Planet Searcher) instrument built for the 3.6 m ESO telescope in Chile (Mayor et al., 2003). Figure 9.2 shows a technical feature of HARPS-N, namely the vacuum-tight and thermostated chamber where the spectrographic module is inserted. Thanks to such innovations, the precision of HARPS-N is well below 1 m/s, dominated by the characteristics of the stellar spectrum, not by instrumental constraints.

The scientific program carried out with HARPS-N at the TNG produces an amazing series of scientific results (e.g. Benatti et al., 2016), and is currently being upgraded in order to provide

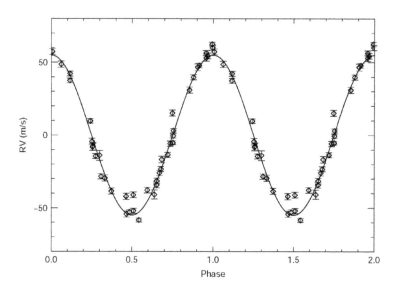

FIGURE 9.1   The radial velocity curve of 51 Pegasi obtained at the TNG with the high-velocity resolution spectrograph SARG. The orbital period of the planet is 4.2 days. (Courtesy R. Gratton, INAF OAPd).

FIGURE 9.2   A detail of the HARPS-N instrument at the TNG inside its vacuum-thermostated chamber. (Courtesy R. Claudi, INAF AOPd.)

simultaneous observations in the visible and near infrared bands (Claudi et al., 2017). Among the highlights of such a program, we recall the first mass estimate for the transiting habitable-zone Super-Earth in a triple system around the star K2-3 of spectral-type M0. Such determination required 329 radial velocity measurements collected over 2.5 years (Damasso et al., 2018).

2. Variation of the luminosity of the star when the planet transits in front of it. Although the amplitude of such variation is exceedingly small, say 1/1,000, modern detectors and good sky allow discovery even by small amateur telescopes. Figure 9.3 shows an example of one such transit, obtained at the TNG with Iqueye, a single-photon detector. The transit lasted about three hours. The amplitude of the dimming of the star was about 0.003. The dispersion of the individual points was mostly due to the atmospheric turbulence.

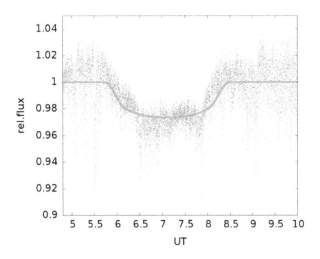

FIGURE 9.3   Transit of an exoplanet over the disk of the star WASP 6. High time resolution data obtained with the fast photometer Iqueye at the TNG. (Courtesy M. Barbieri, INAF AOPd.)

From space, the possibilities are much better, thanks to the lack of atmospheric disturbances, the much darker sky background and the possibility to observe continuously for dozens of hours. Those are the key reasons why NASA's Kepler was so successful. The same principle of detection by luminosity variation has been adopted by the successor of Kepler, namely the Transiting Exoplanet Survey Satellite (TESS, https://tess.gsfc.nasa.gov/), that started operating from a very peculiar orbit in the Earth–Moon system in mid-2018. At the time of writing, TESS has already discovered a super-Earth orbiting the fifth magnitude solar-type star π Mensae, which is about 60 l-y away (Huang et al., 2018). This is the second planet known around this star, the first being a super-Jupiter. Transiting exoplanets can provide a wealth of more information with respect to those who only induce radial velocity variations in their host star because of unfavorable perspective. In particular, their atmospheres can be studied spectroscopically.

The European Space Agency is engaged in exoplanet searches and characterization with three projects:

- CHEOPS (http://sci.esa.int/cheops/) is a small satellite meant to characterize already known exoplanets by very accurate photometry. The launch is foreseen for 2019.

- PLATO (http://sci.esa.int/plato/) is a very large satellite, meant to discover exoplanets with emphasis on the properties of terrestrial planets in the habitable zone around solar-like stars. Some 30 telescopes arranged in a battery will constitute a powerful and photometrically precise measurement system. Its launch toward the outer Lagrangian point L2 of the Sun–Earth system is foreseen for 2024. Why the name PLATO? Because exoplanets are discovered by their shadows, reminding one of the ideas of the great Greek philosopher.

- ARIEL (http://sci.esa.int/ariel), the Atmospheric Remote-Sensing Infrared Exoplanet Large-Survey, will study the chemical composition, the formation and evolution, the thermal structure of exoplanets, by surveying a diverse sample of about 1,000 transiting extra-solar planets, simultaneously in visible and infrared wavelengths. The launch is foreseen for 2028.

9.2 HABITABILITY CONDITIONS OF KNOWN EXOPLANETS

As already remarked in previous chapters, the determination of the conditions for developing and sustaining life on exoplanets is not an easy task. Can cool and small M-stars really support life around them? This is a very debated topic. Dangerous X-rays and UV flares are quite frequent from their chromospheres. In addition, a sufficient supply of a steady UV flux might be necessary for the formation of RNA (Ranjan et al., 2017), and M-stars instead show a UV deficit. With respect to such a deficit, Rimmer et al. (2018) have experimentally estimated the rates of prebiotic photosynthesis in the presence of UV light (a situation they call "light chemistry") versus the rates for the biomolecular reactions that happen in the absence of the UV light ("dark chemistry"). According to the results of this paper, only stars hotter than 4,400 K (K-type stars) can sustain light chemistry. Therefore, the habitability of say Proxima-b and Trappist-1e is by no means certain.

How to recognize biosignatures in exoplanets in light of limited data and interpretation errors in them is discussed in the paper by Kiang et al. (2018). The next generation large telescopes, both on ground and space, will be of paramount importance because (apart from a scanty number of meteorites and some dust of interstellar origin) light is all we have to search for life outside our Solar System. The biosignatures present on the atmosphere or surface of alien worlds may differ significantly from Earth's, simply because of the different spectral type of the illuminating star or may be due to the tidally locked co-rotation of the planet, which might always show the same hemisphere to its nearby star. Oxygen provides an example of possible false detection. Oxygen, produced by photosynthetic organisms on Earth, is a most promising biosigna-ture, but it is not fool-proof. Abiotic processes on a planet could also generate oxygen. Conversely, a planet lacking detectable levels of oxygen could still support life, which was the case of Earth before the global accumulation of oxygen in the atmosphere. A number of life markers must be identified, thanks to specifically designed instruments, before one can be reasonably sure to have found extra-terrestrial life. See the discussion in https://www.jpl.nasa.gov/news/news.php?feature=7171&utm_source=iContact&utm_medium=email&utm_campaign=NASAJPL&utm_content=daily20180625-2.

To close, in this paragraph we mention that the possibility to export life from one planet to another in a multi-planet exo-system, in other words, the possibility of an interplanetary pan-spermia in alien worlds, has been examined by Veras et al. (2018). Can moons of giant gaseous exoplanets support life? While not a candidate for life themselves, Jupiter-like planets in the habitable zone may harbor rocky exo-moons that might provide a favorable environment for life, perhaps even better than Earth, because they receive energy not only from their star but also from radiation reflected from their planet (Hill et al., 2018).

NOTES

1. A coronagraphic device is able to suppress the light of the bright star and leave unaltered the image of the planet, which is typically a fraction of arc seconds away. All modern large telescopes have implemented this facility in their complement of instruments. To be sure, such devices require a great amount of innovative optics, mechanics and controls, both to suppress the light on axis and the atmospheric turbulence when the telescope is on ground and not in space.

2. There are other methods, like gravitational lensing of the OGLE program (see Udalski et al., 2015), not quoted in this Review because they are too specialized. GAIA's increasing importance due to the exploitation of astrometric perturbations has been already quoted.

CHAPTER 10

Search for Extra-Terrestrial Intelligence, SETI

THE STEP FROM "HABITABLE" to "inhabited" is truly a gigantic one, but even more so is the one from "inhabited" to "inhabited by intelligent beings." Many questions can be asked about alien intelligent beings, such as

- How many are there and where are they? See below on the so-called "Fermi paradox."

- How do they communicate?

- What are the social, economic, philosophic and even religious implications of their discovery?

It is said that the Italian physicist Enrico Fermi, during a coffee table conversation asked a question reported in different ways, but whose essence was, "Where are they all? If they exist, why are they not here yet?" Indeed, Carl Sagan quickly showed, solving an equation similar to that of neutron diffusion inside reactors, that an advanced civilization could spread its presence to the whole Milky Way in a few million years. To quantify their number, Frank Drake wrote a simple equation, some 30 years before the discovery of the first exoplanet:

$$N = R_s \times f_P \times n_T \times f_v \times f_i \times f_t \times L \geq 1$$

where:

R_s = the rhythm of formation of suitable stars
f_p = the fraction of stars with habitable planets
n_T = the number of Earth-like planets
$f_v \times f_i \times f_t$ = the fraction of "earths" with life, intelligence and suitable technology
L = the time during which the civilization can communicate,

and where the term ≥ 1 expresses the hope that we too are an intelligent species.

The original version of the equation has been reformulated and discussed several times. We quote two examples, an optimistic one and a pessimistic one.

- Frank and Sullivan (2016) discuss an empirical constraint on the prevalence of technological species in the Universe. They set a lower bound on the probability that one or more technological species have evolved anywhere and at any time in the history of the observable Universe. They find that as long as the probability that a habitable zone planet develops a technological species is larger than $\sim 10^{-24}$, humanity is not the only technological intelligence.

- On a more pessimistic side, Sandberg et al. (2018) recast the equation to represent realistic distributions of uncertainty and found a substantial *ex ante* probability of there being no other intelligent life in our observable universe.

From the large spread of outcomes, it can be questioned if Drake's equation is indeed an equation. Surely, it had and still has the ability to identify the many unknowns in the search for alien intelligence.

Let us consider here the optimistic approach: suppose there is a number of intelligent beings around us. How many of them can we contact? Such a question has been addressed, for instance by Balbi (2018). If the civilizations are located in our Galaxy, the detectability requirement imposes a strict constraint on their epoch of appearance and their communicating lifespan. This, in turn, implies that the fraction of civilizations of which we can find any empirical evidence strongly depends on the specific features of their temporal distribution. A large set of solutions is therefore possible, but the probability of encountering intelligent aliens inside a sphere of 1,000 l-y is typically much less than 1.

Another approach has been taken by Lingam and Loeb (2019), estimating the likelihood of detecting life via technosignatures or via biosignatures. The authors estimate that the likelihood of detecting intelligent life might be two orders of magnitude smaller compared to the detection of primitive life. It should be noted that this estimate can be seen as an upper bound since it will be lowered significantly if technological intelligence is rare or short-lived.

Surely, finding alien civilization is a very difficult but not hopeless undertaking. There is even an Institute in the USA carrying the SETI name (https://www.seti.org/). Their website is full of relevant information and is well worth reading.

Realistically, we can think of several strategies in order to find alien civilizations:

1. Improve our telescopes and detection devices and strategies

2. Listen for intelligent signals, in the radio or optical domain

3. Signal our existence and let them find us

Regarding point 1, many very large optical telescopes are under construction, like the 39m European Large Telescope in Chile (https://www.eso.org/sci/facilities/eelt/). The launch of the NASA 6.5m James Web Space Telescope (https://www.jwst.nasa.gov/) to the L2 point of the Sun-Earth system has been delayed to 2021, but it will eventually see its first light. In the radio domain, in addition to the already quoted FAST, we cannot forget the Square Kilometer Array (SKA, https://www.skatelescope.org/), a gigantic scientific, technological and even political effort because many nations worldwide are involved in such venture. The SKA will use thousands of dishes and up to a million low-frequency antennas that will enable observers to monitor the sky in unprecedented detail and survey the entire sky much faster than any system currently in existence.

Therefore, we can be confident in a large step forward for astronomical devices around the end of the next decade.

Listening to radio signals (point 2) is the most promising avenue, pursued since the very early days of SETI. Many projects explore the so-called "water hole" band between 1.1 to 1.9 GHz, where the noise of natural origin is at its minimum. Too bad our own signals contaminate such a band in a steadily increasing way.

Among the several initiatives currently underway, we quote the Berkeley Breakthrough Listen Initiative (http://astro.berkeley.edu/p/breakthrough-listen), which has already released a fair amount of data obtained by listening to 700 or so stars, with no sign of artificial signals. Other projects are carried out at Parkes in Australia. In Italy, a radio SETI project is managed by the INAF station in Medicina, near Bologna (https://www.ira.inaf.it/Home.html).

Optical SETI is pursued in rare attempts; see for instance http://seti.harvard.edu/oseti/, but again, there has been no success until now.

The failures might raise the interesting question of whether alien civilizations wish to conceal their presence by using quantum technologies for their communications. Our capabilities in such a novel field are rapidly expanding, both on terrestrial stations and space satellites, so that in the near future we would be able to detect alien quantum signals.

Regarding point 3, the idea to signal our existence and "let them find us" might be scary to many people. Attempts have been made, starting from the bold ideas of the golden records on the Pioneer and Voyager spacecrafts conceived by Carl Sagan, and the digital signal sent by the Arecibo radiotelescope around 1970 toward the remote globular cluster M13. Such an attempt at signal transmission has been repeated very recently by sending text and music toward the exoplanet of the nearby star Gliese 272, a mere 12 light-years away (see http://meti.org/mission, where the *m* in *meti* means "messaging"). If anybody there listens and answers, in two dozen years we will receive their reply.

Indeed, we are already signaling our existence, even if in an unwanted manner, through huge and constantly increasing amounts of electromagnetic waves which Earth emits in the forms of artificial illumination and radio broadcasts. Take into account that nearby alien astronomers (assuming they exist) will be lead naturally to intensively study our Sun, which is by no means an average star, as is commonly said. First of all, it is a single well isolated star, while the great majority of stars are in double or even more complex systems and inside clusters. Second, its luminosity places it in approximately the upper 5% of the most luminous stars, its temperature is ideal for optical studies, and it is fairly stable in such a band over millennia. Moreover, centuries of observations will reveal the transit of an amazing set of planets (not very different, apart from the spatial scale, from that of Trappist-1), some of them with atmospheres and magnetic fields as evidenced by aurorae. In the course of the last century, the alien astronomers would have noticed that planet number 3 (Earth) had an incredibly rapid increase of CO_2 and artificial substances containing fluorine and chlorine.

Moreover, the transits of the planet are becoming less and less "dark." To be specific, let us examine the optical band (see

Figure 10.1), during the 14-hour transit of the planet over the stellar disk (our alien colleagues are smart enough to remove occasional disturbances due to clouds).

The alien astronomers have enough evidence to conclude that Earth is inhabited by a technologically advanced civilization, capable of launching a plethora of chemically fueled rockets. Therefore, they would conclude that it is well worth sending spacecraft to explore such a peculiar planet, e.g., by sending very light cameras attached to huge sails pushed by powerful lasers.

Coming back to our situation, it looks like we could do the same with present technologies, given enough resources. See for instance Heller et al. (2017). The authors calculate that although Alpha Centauri, which contains the planet Proxima-b, is the most nearby star system, Sirius offers the shortest possible travel times in a bound orbit: 69 years assuming 12.5% of the velocity of light can be obtained at departure from the Solar System.

Of course, those are dreams of a very distant future. At the moment of writing, there is absolutely no sign of "their" existence. The future is surely unpredictable, so we cannot stop the search.

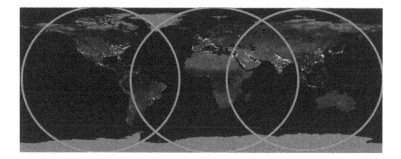

FIGURE 10.1   Artificial illumination of Earth (situation around 2015). The circles indicate the 14 hours of the transit of Earth over the solar disk. Background image credit NASA.

If contact will be established, or even if they come here, the consequences on society are difficult to predict, simply judging from the impressive variety of opinions expressed in the literature. Fear and hope are more or less equally distributed. Scientific, technological and even religious matters will be under fierce discussion (Barbieri, 2000; Davies, 1995; Consolmagno and Mueller, 2014).

CHAPTER 11

Conclusions

T HE BUILDING BLOCKS OF life are present everywhere, from Earth, the planets and their moons to asteroids and comets, to exoplanets. However, the crucial passage from those elements to life is still a mystery. Biologists, chemists and physicists are tackling the study of this crucial step with great energy. Astronomers can help them by pointing out where and how to search for and maybe discover extra-terrestrial life. It might well be that only the discovery of extra-terrestrial life will provide clues to dissipate this mystery.

In this astronomically oriented review, we have underlined that the theme of life in extra-terrestrial environments has different faces. Microbial life may have been present, or may even be present today, in several bodies of our own Solar System; ground telescopes and space missions are vigorously searching for it. In the near future, we might export our own life to permanent bases on the Moon and Mars, overcoming the many risks and challenges of such a gigantic enterprise. Ground and space telescopes, laboratory experiments and theoretical models are continuously improving our understanding of the conditions of habitability on exoplanets, while the search for other civilizations somewhere in the Universe is carried out by radio and optical means.

The outcome of the great variety of attempts to search for extra-terrestrial life is not easily foreseeable. Therefore, the sentence of Galileo Galilei, that "the Book of Nature is written in the language of mathematics" (and we might add today the languages of physics, chemistry and biology), is as true as ever. It is our duty to read the book of nature with the greatest freedom of mentality.

As Shakespeare wrote around 1602, when Galileo and Kepler opened the way to modern astronomy:

> *Hamlet: There are more things in heaven and earth, Horatio, than are dreamt of in your philosophy.*

Indeed, we astronomers cannot cease to observe the sky, not only with the hope but also with the certainty that we will make more discoveries.

References

Abramov O and Mojzsis SJ (2009) Microbial habitability of the Hadean Earth during the late heavy bombardment, *Nature*, Vol. 459, Issue 7245, pp. 419–422. DOI: 10.1038/nature08015.

Agarwal J et al. (2017) Evidence of sub-surface energy storage in comet 67P from the outburst of 2016 July 03, *Monthly Notices of the Royal Astronomical Society*, Vol. 469, Issue Suppl_2, pp. s606–s625. DOI: 10.1093/mnras/stx2386.

Ajello M and the Fermi-LAT Collaboration (2018) A gamma-ray determination of the Universe's star formation history, *Science*, Vol. 362, Issue 6418, pp. 1031–1034. DOI: 10.1126/science.aat8123.

Altwegg K et al. (2016) Prebiotic chemicals—Amino acids and phosphorous—in the coma of comet 67P/Churyumov-Gerasimenko. *Science Advances*, Vol. 2016, Issue 2, e1600285. DOI: 10.1126/sciadv.1600285.

Altwegg K et al. (2017) D_2O and HDS in the coma of 67P/Churyumov-Gerasimenko, *Philosophical Transactions of the Royal Society A*, Vol. 375, Issue 2097, id.20160253. DOI: 10.1098/rsta.2016.0253.

Asplund M, Grevesse N, Sauval AJ and Scott P (2009) The chemical composition of the Sun, *Annual Review of Astronomy and Astrophysics*, Vol. 47, Issue 1, pp. 481–522. DOI: 10.1146/annurev.astro.46.060407.145222.

Bailer-Jones AL et al. (2018) Plausible home stars of the interstellar object 'Oumuamua found in Gaia DR2, arXiv:1809.09009v1 [astro-ph. EP], accepted to *The Astronomical Journal*.

Balbi A (2018) the impact of the temporal distribution of communicating civilizations on their detectability, *Astrobiology*, Vol. 18, Isssue 1. DOI: 10.1089/ast.2017.1652.

Ball P (2018) What is life? *Nature*, Vol. 560, pp. 548–550.

Bandfield JL, Poston MJ, Klima RL and Edwards CS (2018) Widespread distribution of OH/H$_2$O on the lunar surface inferred from spectral data, *Nature Geoscience*, DOI: 10.1038/s41561-018-0065-0.

Barbieri C (2000) L'influenza del Cristianesimo sullo sviluppo dell'Astronomia, in *Dopo 2000 anni di Cristianesimo*, S. Romano ed., Mondadori, Milan, Italy.

Barbieri C et al. (2002) PLEXISS: a coronagraph for imaging the lunar atmosphere from the International Space Station, Proceedings of the SPIE, Vol. 4767, pp. 106–113. DOI: 10.1117/12.451223.

Barbieri C (2017) Comet 67P/C-G seen through Osiris, the eyes of Rosetta, Rendiconti Lincei. *Scienze Fisiche e Naturali*, Vol. 28, Issue 9. DOI: 10.1007/s12210-017-0618-y.

Barbieri C and Bertini I (2017) Comets, *Rivista del Nuovo Cimento Issue B August*, pp. 335–409. DOI: 10.1393/ncr/i2017-10138-4.

Barge LM and White LM (2017) Experimentally testing hydrothermal vent origin of life on enceladus and other icy/ocean worlds, *Astrobiology* Vol. 17, Issue 9, DOI: 10.1089/ast.2016.1633.

Benatti S et al. (2016) The GAPS Project: First Results, *Frontier Research in Astrophysics II*, held 23–28 May, 2016 in Mondello (Palermo), Italy. Online at https://pos.sissa.it/cgi-bin/reader/conf.cgi?confid=269, id.69.

Bennett J and Shostak S (2007) *Life in the Universe*, Pearson Addison Wesley, San Francisco, CA.

Bibring JP et al. (2006) Global mineralogical and aqueous mars history derived from OMEGA/mars express data, *Science 21*, Vol. 312, Issue 5772, pp. 400–404. DOI: 10.1126/science.1122659.

Bobroskiy I et al. (2018) Ancient steroids establish the Ediacaran fossil Dickinsonia as one of the earliest animals, *Science* Vol. 361, Issue 6408, pp. 1246–1249. DOI: 10.1126/science.aat7228.

Bockelée-Morvan D et al. (2015) Cometary isotopic measurements, *Space Science Reviews*, Vol. 197, pp. 47–83. DOI: 10.1007/s11214-015-0156-9.

Bonnet RM and Woltjer L (2008) *Surviving 1,000 Centuries: Can We Do It?*, Springer, Berlin.

Bottke, WF and Norman MD (2017) The late heavy bombardment, Annual *Review of Earth and Planetary Sciences*, Vol. 45, pp. 619–647, DOI: 10.1146/annurev-earth-063016-020131.

Camuffo, D (2000) Lunar influence on climate, in *Earth-Moon Relationship*, Barbieri C and Rampazzi F eds., pp. 99–113, Kluwer Academic Publishers, Dordrecht, the Netherlands.

Canup R (2004) Dynamics of lunar formation, *Annual Review of Astronomy &Astrophysics*, Vol. 42, Issue 1, pp. 441–475. DOI: 10.1146/annurev.astro.41.082201.113457.

Capova KA, Persson E, Milligan T and Dunér D eds. (2018) *Astrobiology and Society in Europe Today*, Springer, Berlin.

Castaldo L, Mège D, Gurgurewicz J, Orosei R and Alberti G (2017) Global permittivity mapping of the Martian surface from SHARAD, *Earth and Planetary Science Letters*, Vol. 462, pp. 55–65, DOI: 10.1016/j.epsl.2017.01.012.

Ceccarelli C et al. (2017) Seeds of life in space (SOLIS): The organic composition diversity at 300–1000 au scale in solar-type star forming regions, *The Astrophysical Journal*, Vol. 850, Issue 2. DOI: 10.3847/1538-4357/aa961d.

Chou Yu-Min et al. (2018) Multidecadally resolved polarity oscillations during a geomagnetic excursion, *Proceedings of the National Academy of Sciences of the United States of America* DOI: 10.1073/pnas.1720404115.

Ciarletti V (2016 A variety of radars designed to explore the hidden structures and properties of the solar system's planets and bodies, *Comptes Rendus Physique*, Vol. 17, Issue 9, pp. 966–975.

Claudi R. et al. (2017) GIARPS@TNG: GIANO-B and HARPS-N together for a wider wavelength range spectroscopy, *European Physical Journal Plus*, Vol. 132, 364. DOI: 10.1140/epjp/i2017-11647-9.

Cockell CS (2015) *Astrobiology: Understanding Life in the Universe*, John Wiley & Sons, Chichester.

Cockell CS (2017) The laws of life, *Physics Today*, Vol. 70, Issue 3, 42. DOI: 10.1063/PT.3.3493.

Consolmagno GSJ and Mueller PSJ (2014) *Would You Baptize an Extraterrestrial?*, Crown Publishing Group, New York.

Cosentino R et al. (2012) Harps-N: the new planet hunter at TNG, *SPIE*, Vol. 8446, Ground-based and Airborne Instrumentation for Astronomy IV, 84461V, DOI: 10.1117/12.925738.

Cottin H et al. (2017a) Space as a tool for astrobiology: Review and recommendations for experimentations in earth orbit and beyond, *Space Science Reviews*, Vol. 209, Issue 1–4, pp. 83–181.

Cottin H et al. (2017b) *Comets and Astrobiology, (Re)Assessment for Comet 67P After Rosetta, XVIIIth International Conference on the Origin of Life, Proceedings of the conference held 16-21 July, 2017 in San Diego, California*. LPI Contribution No. 1967, 2017, id.4082, 2017LPICo1967.4082C.

Courde et al, (2017) Lunar laser ranging in infrared at the Grasse laser station, *Astronomy & Astrophysics*, Vol. 602, id.A90, DOI: 10.1051/0004-6361/201628590.

Covino S et al. (2017) The unpolarized macronova associated with the gravitational wave event GW 170817, *Nature Astronomy*, Vol. 1, pp. 791–794. DOI: 10.1038/s41550-017-0285-z.

Cramwinckel MJ et al. (2018) Synchronous tropical and polar temperature evolution in the Eocene, *Nature*, Vol. 559, pp. 382–386.

Damasso M et al. (2018) Eyes on K2-3: A system of three likely sub-Neptunes characterized with HARPS-N and HARPS, *Astronomy & Astrophysics*, Vol. 615, id.A69, DOI: 10.1051/0004-6361/201732459.

Davidsson B et al. (2016) The primordial nucleus of comet 67P/Churyumov-Gerasimenko, *Astronomy & Astrophysics*, Vol. 592, id.A63, p. 30. DOI: 10.1051/0004-6361/201526968.

Davies, P (1995) *Are We Alone?*, Basic Books, New York.

De Angelis A and Pimenta M (2018) *Introduction to Particle and Astroparticle Physics Multimessenger Astronomy and its Particle Physics Foundations*, 2nd edition, Springer, Berlin.

de Bergh C, Lutz BL, Owen T and Chauville J (1988) Monodeuterated methane in the outer solar system. III—Its abundance of Titan, *Astrophysical Journal, Part 1*, Vol. 329, pp. 951–955. DOI: 10.1086/166439.

Del Genio AD, Brain D, Noack L and Schaefer L (2018) *Divergent Climate and Habitability Histories*. https://arxiv.org/abs/1807.04776.

Di Mauro E and Saladino F. (2016) *Dal Big Bang alla cellula madre, L'origine della vita*, Casa Editrice Il Mulino, Bologna, Italy.

Doglioni C and Panza G (2015) Polarized plate tectonics, Advances in Geophysics, Vol. 56, Issue 3, pp. 1–167. DOI: 10.1016/bs.agph.2014.12.001.

Etiope G (2018) Understanding the origin of methane on mars through isotopic and molecular data from the ExoMars orbiter, *Planetary and Space Science*, Vol. 159, pp. 93–96.

Favre C et al. (2018) First detection of the simplest organic acid in a protoplanetary disk, *ApJL*, Vol. 862, L2. DOI: 10.3847/2041-8213/aad046.

Fitzgerald RJ (2018) New constraints on early oxygen levels, *Physics Today*. DOI: 10.1063/PT.6.1.20180725a.

Frank A and Sullivan WT (2016) A new empirical constraint on the prevalence of technological species in the Universe, *Astrobiology*, Vol. 16, Issue (5), pp. 359–62. DOI: 10.1089/ast.2015.

Frankel HR (2012) *The Continental Drift Controversy*, four volumes, Cambridge University Press, Cambridge.

Gasda P et al. (2017) In situ detection of boron by ChemCam on Mars, *Geophysical Research Letters*, DOI: 10.1002/2017GL074480.

Geiss J and Reeves H (1972) Cosmic and solar system abundances of Deuterium and Helium-3, *Astron. & Astrophys.*, Vol. 18, pp. 126–132.

Gibney E (2018) How to build a Moon base, *Nature*, Vol. 562, pp. 474–478.

Gillon M et al. (2017) Seven temperate terrestrial planets around the nearby ultracool dwarf star TRAPPIST-1, *Nature*, Vol. 542, pp. 456–460. DOI: 10.1038/nature21360.

Glavin DP et al. (2012) Unusual nonterrestrial L-proteinogenic amino acid excesses in the Tagish Lake meteorite, *Meteoritics & Planetary Science*, Vol. 47, Issue 8, pp. 1347–1364. DOI:10.1111/j.1945-5100.2012.01400.x.

Goldsmith D and Owen T (2002) *The Search of Life in the Universe*, University Science Books, Sausalito, CA.

Guijarro A and Yurs M (2007) *The Origin of Chirality in the Molecules of Life*, Royal Society of Chemistry Publishing, Cambridge, UK.

Hallis LJ (2017) D/H ratios of the inner solar system, *Philosophical Transactions of The Royal Society A Mathematical Physical and Engineering Sciences*, Vol. 375, Issue 2094, id.20150390. DOI: 10.1098/rsta.2015.0390.

Hashimoto T et al. (2018) The onset of star formation 250 million years after the Big Bang, *Nature*, Vol. 557, pp. 392–395.

Heller E, Hippke M and Kervella P (2017) Optimized trajectories to the nearest stars using lightweight high-velocity photon sails, *Astronomical Journal*, Vol. 154, 115. DOI: 10.3847/1538-3881/aa813f.

Helmi A et al. (2018) The merger that led to the formation of the Milky Way's inner stellar halo and thick disk, *Nature*, Vol. 563, pp. 85–88.

Hemingway DJ and Tikoo SM (2018) Lunar Swirl morphology constrains the geometry, magnetization, and origins of lunar magnetic anomalies *Journal of Geophysical Research Planets*, Vol. 123, Issue 8. DOI: 10.1029/2018JE005604.

Hill M et al. (2018) Exploring Kepler giant planets in the habitable zone, *The Astrophysical Journal*, Vol. 860, 1. DOI: 10.3847/1538-4357/aac384.

Huang CX et al. (2018) TESS discovery of a transiting Super-Earth in the Π Mensae system, arXiv:1809.05967v1 [astro-ph.EP].

Jakosky B et al. (2018a) Loss of the Martian atmosphere to space: Present-day loss rates determined from MAVEN observations and integrated loss through time, *Icarus*, Vol. 315, pp. 146–157. DOI: 10.1016/j.icarus.2018.05.030.

Jakosky B et al. (2018b) *Inventory of CO_2* available for terraforming mars, *Nature Astronomy*, Vol. 2, pp. 634–639. DOI: 10.1038/s41550-018-0529-6.

Jia X, Kivelson MJ, Khurana KK and Kurth WS (2018) Evidence of a plume on Europa from Galileo magnetic and plasma wave signatures, *Nature Astronomy*, Vol. 2, pp. 459–464.

Johns D et al. (2018) Revised exoplanet radii and habitability using Gaia Data Release 2, https://arxiv.org/abs/1808.04533.

Kiang NY et al. (2018) Exoplanet biosignatures: At the dawn of a new era of planetary observations, *Astrobiology*, Vol. 18, Issue 6. DOI: 10.1089/ast.2018.1862.

Kim SC, O'Flaherty DK, Zhou L, Lelyveld VS and Szostak JW (2018) Inosine, but none of the 8-oxo-purines, is a plausible component of a primordial version of RNA, *Proceedings of the National Academy of Sciences*, Vol. 115, Issue 52, pp. 13318–13323. DOI: 10.1073/pnas.1814367115.

Koehler MC et al. (2018) Transient surface ocean oxygenation recorded in the ~2.66-Ga Jeerinah Formation, Australia, *Proceedings of the National Academy of Sciences* http://www.pnas.org/content/early/20.

Koop L et al. (2018) High early solar activity inferred from helium and neon excesses in the oldest meteorite inclusions, *Nature Astronomy*, Vol. 2, pp. 709–713. DOI: 10.1038/s41550-018-0527-8.

Laskar J et al. (2004a) A long-term numerical solution for the insolation quantities of the Earth, *Astrophysics & Astronomy*, Vol. 428, pp. 261–285. DOI: 10.1051/0004-6361:20041335.

Laskar J et al. (2004b) Long-term evolution and chaotic diffusion of the insolation quantities of Mars, *Icarus*, Vol. 170, Issue 2, pp. 343–364. DOI: 10.1016/j.icarus.2004.04.005.

Laskar J et al. (2011) La2010: a new orbital solution for the long-term motion of the Earth, *Astrophysics & Astronomy*, Vol. 532, p. A89. DOI: 10.1051/0004-6361/201116836.

Lawver LA, Dalziel IWD, Norton IO and Gahagan LM (2009) *The PLATES 2009 Atlas of Plate Reconstructions (750 Ma to Present Day)*, PLATES Progress Report No. 325-0509, University of Texas Technical Report No. 196, p. 57.

Levison F, Morbidelli A, Vanlaerhoven C, Gomes R, and Tsiganis K (2008) Origin of the structure of the Kuiper belt during a dynamical instability in the orbits of Uranus and Neptune, *Icarus*, Vol. 196, Issue 1, pp. 258–273.

Li S et al. (2018) Direct evidence of surface exposed water ice in the lunar polar regions, *Proceedings of the National Academy of Sciences*, www.pnas.org/cgi/doi/10.1073/pnas.1802345115.

Lingam M and Loeb A (2019) Relative Likelihood of Success in the Searches for Primitive versus Intelligent Extraterrestrial Life, *Astrobiology*, Vol. 19, Issue 1, pp. 28–39. DOI: 10.1089/ast.2018.1936.

Livio M (2018) *How Special Is the Solar System?*, chapter of the book *Consolidation of Fine Tuning*, eprint arXiv:1801.05061.

Lock SJ et al. (2018) The origin of the Moon within a terrestrial synestia, *Journal of Geophysical Research: Planets*, Vol. 123, Issue 4, pp. 910–951. https://arxiv.org/abs/1802.10223.

Lucchetti A, Pozzobon R, Mazzarini F, Cremonese G and Massironi M (2017) Brittle ice shell thickness of Enceladus from fracture distribution analysis, *Icarus*, Vol. 297, pp. 252–264.

Lyons TW, Droser ML, Lau KV and Porter SM (2018) Early Earth and the rise of complex life, *Emerging Topics in Life Sciences*, Vol. 2, Issue 2, pp. 121–124. DOI: 10.1042/ETLS20180093.

Marchi S et al. (2018) An aqueously altered carbon-rich Ceres, *Nature Astronomy Letters*, Vol. 3, pp. 140–145. DOI: 10.1038/s41550-018-0656-0.

Marounina N, Grasset R, Tobie G and Carpy S (2018) Role of the global water ocean on the evolution of Titan's primitive atmosphere, *Icarus*, Vol. 310, pp. 127–139. DOI: 10.1016/j.icarus.2017.10.048.

Martins Z et al. (2017) Earth as a tool for astrobiology—A European perspective, *Space Science Reviews*, Vol. 209, Issue 1–4, pp. 43–81. DOI: 10.1007/s11214-017-0369-1.

Mayor M and Queloz D (1995) A Jupiter-mass companion to a solar-type star, *Nature*, Vol. 378, Issue 6555, pp. 355–359, DOI: 10.1038/378355a0.

Mayor M et al. (2003) Setting new standards with HARPS, *The Messenger*, Issue 114, pp. 20–24.

McGuire B et al. (2016) Discovery of the interstellar chiral molecule propylene oxide (CH_3CHCH_2O), *Science*, Vol. 352, Issue 6292, pp. 1449–1452. DOI: 10.1126/science.aae0328.

McGuire B et al. (2018) *First Results of an ALMA Band 10 Spectral Line Survey of NGC 6334I: Detections of Glycolaldehyde (HC(O) CH2OH) and a New Compact Bipolar Outflow in HDO and CS*, https://arxiv.org/pdf/1808.05438.pdf, accepted to ApJ letters.

Meier MMM and Alwmark SH (2017) A tale of clusters: no resolvable periodicity in the terrestrial impact cratering record, *Monthly*

Notices of the Royal Astronomical Society, Vol. 467, Issue 3, pp. 2545–2551.

Metzger PT, Sykes MV, Stern A and Runyon K (2018) *The Reclassification of Asteroids from Planets to Non-Planets*. https://arxiv.org/abs/1805.04115v2.

Michael G, Basilevsky A and Neukum G (2018) On the history of the early meteoritic bombardment of the Moon: Was there a terminal lunar cataclysm? *Icarus*, Vol. 302, pp. 80–103. DOI: 10.1016/j.icarus.2017.10.046.

Micheli M et al. (2018) Non-gravitational acceleration in the trajectory of 1I/2017 U1, *Nature*, Vol. 559, 222. DOI: 10.1038/s41586-018-0254-4.

Milankovitch M. (1930) *Mathematische Klimalhre und Astronomische Theorie der Klimaschwankungen*, Gebruder Borntraeger, Berlin.

Miller J (2018) Isotope measurements help pin down the ancient rise of oxygen, *Physics Today*. DOI: 10.1063/PT.3.3939.

Millot M et al. (2018) Experimental evidence for superionic water ice using shock compression, *Nature Physics*. DOI:10.1038/s41567-017-0017-4.

Morbidelli A et al. (2018) The timeline of the Lunar bombardment—revisited. *Icarus*, Vol. 305, pp. 262–276. DOI: 10.1016/j.icarus.2017.12.046.

Naylor E (2000) Marine annual behavior in relation to lunar phases, in *Earth-Moon Relationship*, Barbieri C and Rampazzi F eds. pp. 291–302, Kluwer Academic Publishers, Dordrecht, the Netherlands.

Nestola F and Smyth JR (2016) Diamonds and water in the deep Earth: a new scenario, International Geology Review, Vol. 58, Issue 3, pp. 263–276. DOI: 10.1080/00206814.2015.1056758.

Noguchi M (2018) The formation of solar-neighbourhood stars in two generations separated by 5 billion years, *Nature*, Vol. 559, pp. 585–588.

Oba Y, Watanabe N, Osamura Y and Kouchi A (2015) Chiral glycine formation on cold interstellar grains by quantum tunneling hydrogen-deuterium substitution reactions, *Chemical Physics Letter*, Vol. 634, pp. 53–59.

Ohishi M, Suzuki T, Hirota T, Saito M and Kaifu N (2017) *Detection of New Methylamine (CH3NH2) Sources: Candidates for Future Glycine Surveys*, eprint arXiv:1708.06871.

Oliveira JS and Wieczorek MA (2017) Testing the axial dipole hypothesis for the Moon by modeling the direction of crustal magnetization,

Journal of Geophysical Research: Planets, Vol. 122, Issue 2, pp. 383–399. DOI: 10.1002/2016JE005199.

Orosei R. et al. (2018) Radar evidence of subglacial liquid water on Mars, *Science*, Vol. 361, Issue 6401. DOI: 10.1126/science.aar7268.

Owen T, Lutz BL and de Bergh C (1986) Deuterium in the outer solar system—Evidence for two distinct reservoirs, *Nature*, Vol. 320, pp. 244–246. DOI: 10.1038/320244a0.

Pearce BKD, Pudritz, Ralph E, Semenov, Dmitry A and Hennin, Thomas K (2018a) Origin of the RNA world: the fate of nucleobases in warm little ponds, *Proceedings of the National Academy of Sciences of the United States of America*, Vol. 1114, Issue 43, pp. 11327–11332. DOI: 10.1073/pnas.1710339114.

Pearce BKD, Tupper, Andrew S, Pudritz, Ralph E and Higgs, Paul G. (2018b) Constraining the time interval for the origin of life on earth *Astrobiology*, Vol. 18, pp. 343–364. DOI: 10.1089/ast.2017.1674.

Pfalzner S, Bhandare A, Vincke K and Lacerda P (2018) Outer solar system possibly shaped by a stellar fly-by, *The Astrophysical Journal*, Vol. 863, Issue 1, id. 45. DOI: 10.3847/1538-4357/aad23c.

Pian P. et al. (2017) Spectroscopic identification of r-process nucleosynthesis in a double neutron star merger, *Nature*, Vol. 551, Issue 7678, pp. 67–70. DOI: 10.1038/nature24298.

Postberg F et al. (2018) Macromolecular organic compounds from the depths of Enceladus, *Nature*, Vol. 558, pp. 564–568.

Qin C, Zhong S and Phillips R (2018). Formation of the lunar fossil bulges and its implication for the early Earth and Moon. *Geophysical Research Letters*, Vol. 45. DOI: 10.1002/2017GL076278.

Ramirez RM (2018) Invited review for publication in *Planetary Evolution and Search for Life on Habitable Planets*, Special Issue (58 pages, 15 Figures, 1 Table) https://arxiv.org/ftp/arxiv/papers/1807/1807.09504.pdf.

Ranjan S, Wordsworth R and Sasselov D (2017) The surface UV environment on planets orbiting M dwarfs: implications for prebiotic chemistry and the need for experimental follow-up, *The Astrophysical Journal*, Vol. 843, Issue 2, article id. 110. DOI: 10.3847/1538-4357/aa773e.

Ribas I et al. (2018) A candidate super-Earth planet orbiting near the snow line of Barnard's star, *Nature*, Vol. 563 pp. 365–368. DOI: 10.1038/s41586-018-0677-y.

Rice TS, Bergin EA, Jørgensen JS and Wampfler SF (2018) Exploring the Origins of Earth's Nitrogen: Astronomical Observations of Nitrogen-bearing Organics in Protostellar Environments, *The*

Astrophysical Journal, Vol. 866, Issue 2, 156. DOI: 10.3847/1538-4357/aadfdb.

Rickman et al. (2017) Cometary impact rates on the Moon and planets during the late heavy bombardment, *Astronomy & Astrophysics*, Vol. 598, id. A67. DOI: 10.1051/0004-6361/201629376.

Rimmer P et al. (2018) The Origin of RNA Precursors on Exoplanets, *Science Advances*, Vol. 4, Issue 8, eaar3302. DOI: 10.1126/sciadv.aar3302.

Sandberg A, Eric Drexler and Toby Ord (2018) *Dissolving the Fermi Paradox*, arXiv:1806.02404v1 [physics.pop-ph].

Saur J et al. (2015) The search for a subsurface ocean in Ganymede with Hubble Space Telescope observations of its auroral ovals, *JGR*, Vol. 120, Issue 3, pp. 1715–1737. DOI: 10.1002/2014JA020778.

Scafetta N, Milani F, Bianchini A and Ortolani S (2016) On the astronomical origin of the Hallstatt oscillation found in radiocarbon and climate records throughout the Holocene, *Earth Science Review*, Vol. 162, pp. 24–43.

Schad W (2000) Lunar influence on plants, in *Earth-Moon Relationship*, Barbieri C and Rampazzi F eds. pp. 405–409, Kluwer Academic Publishers, Dordrecht, the Netherlands.

Schenk PM et al. (2018a) Basins, fractures and volcanoes: Global cartography and topography of Pluto from New Horizons, *Icarus*, Vol. 314, pp. 400–433. DOI: 10.1016/j.icarus.2018.06.008.

Schenk PM et al. (2018b) Breaking up is hard to do: global cartography and topography of Pluto's mid-sized icy Moon Charon from New Horizons, *Icarus*, Vol. 315, pp. 124–145. DOI: 10.1016/j.icarus.2018.06.010.

Schoepp-Cothenet B et al. (2012) The ineluctable requirement for the trans-iron elements molybdenum and/or tungsten in the origin of life, *Scientific Reports*, Vol. 2, p. 263. DOI: 10.1038/srep00263.

Schrödinger E (1944) *What Is Life? The Physical Aspect of the LIving Cell*, Cambridge University Press, Cambridge.

Schulze-Makuch D and Crawford IA (2018) Was There an Early Habitability Window for Earth's Moon? *Astrobiology*, Vol. 18, Issue 8. DOI: 10.1089/ast.2018.1844.

Schwadron NA et al. (2018) Update on the worsening particle radiation environment observed by CRaTER and implications for future human deep-space exploration, *Space Weather*, Vol. 16, Issue 3, DOI: 10.1002/2017SW001803.

Silk J (2018) Put a radiotelescope on the far side of the Moon, *Nature*, Vol. 553, p. 6.

Snellen I and Brown A (2018) The mass of the young planet Beta Pictoris b through the astrometric motion of its host star, *Nature Astronomy*, Vol. 2, pp. 883–886.

Sparks WB et al. (2017) Active cryovolcanism on Europa?. *Astrophysical Journal Letters*, Vol. 839, L18.

Speyerer EJ et al. (2016) Quantifying crater production and regolith overturn on the Moon with temporal imaging, *Nature*, Vol. 538, pp. 215–218.

Spiga R, Barbieri C, Bertini I, Lazzarin M and Nestola F (2019) The origin of water on Earth: stars or diamonds? A conversation among astronomers and geologists, *Rendiconti Accademia dei Lincei, Scienze Fisiche e Naturali*, DOI: 10.1007/s12210-018-0753-0.

Steffen W. et al. (2018) Trajectories of the Earth system in the Anthropocene, *Proceedings of the National Academy Sciences of the United States of America*, Vol. 115, Issue 33, pp. 8252–8259. DOI: 10.1073/pnas.1810141115.

Strom, RG, Marchi S and Malhotra R (2018) Ceres and the terrestrial planets impact cratering record, *Icarus*, Vol. 302, pp. 104–108, DOI: 10.1016/j.icarus.2017.11.013.

Tachibana S et al. (2017) Liquid-like behavior of UV-irradiated interstellar ice analog at low temperatures, *Science Advances*. DOI: 10.1126/sciadv.aao2538.

Terada K, Yokota S, Saito Y, Kitamura N and Asamura Nishino MN (2017) Biogenic oxygen from Earth transported to the Moon by a wind of magnetospheric ions, *Nature Astronomy*, Vol. 1 id. 26. DOI: 10.1038/s41550-016-0026.

Thomas N, et al. (2017) The colour and stereo surface imaging system (CaSSIS) for the ExoMars trace gas orbiter, *Space Science Reviews*, Vol. 212, 1897. DOI: 10.1007/s11214-017-0421-1.

Tsiganis K, Gomes R, Morbidelli A and Levison HF (2005) Origin of the orbital architecture of the giant planets of the solar system, *Nature*, Vol. 435, pp. 459–461, DOI: 10.1038/nature03539.

Udalski A, Szymański MK and Szymański G (2015) OGLE-IV: fourth phase of the optical gravitational lensing experiment, *Acta Astronomica*, Vol. 65, Issue 1, pp. 1–38, arXiv:1504.05966v1 [astro-ph.SR].

Vago L et al. (2017) Habitability on early mars and the search for biosignatures with the ExoMars Rover, *Astrobiology*, Vol. 17, Issue 6–7, pp. 471–510. DOI: 10.1089/ast.2016.1533.

Verani S, Barbieri C, Benn C, Cremonese G and Mendillo M (2001) The 1999 Quadrantids and the lunar Na atmosphere. *Monthly Notices of The Royal Astronomical Society*, Vol. 327, pp. 244–248.

Veras D et al. (2018) Dynamical and biological panspermia constraints within multi-planet exosystems, arXiv:1802.04279v2 [astro-ph. EP].

Walker I, Davies PC and Ellis GFR editors (2017) *From Matter to Life, Information and Causality*, Cambridge University Press, Cambridge, UK.

Ward P. D. and Brownlee D. (2009) *Rare Earth, Why Complex Life Is Uncommon in the Universe*, Copernicus Book, DOI 10.1007/978-0-387-21848-9.

Wegener A (1966) *The Origin of Continents and Oceans* [*Die Entstehung der Kontinente und Ozeane*, 1929], translated from the German by John Biram, Dover Publications, New York.

Williams RJP and Fausto da Silva JJR (2002) The involvement of molybdenum in life, *Biochemical and Biophysical Research Communications*, Vol. 292, Issue 2, pp. 293–299. DOI: 10.1006/bbrc.2002.6518.

Wilson JK, Mendillo M and Spence HE (2006) Magnetospheric influence on the Moon's exosphere, *Journal of Geophysical Research: Space Physics*, Vol. 111, Issue A7. DOI: 10.1029/2005JA011364.

Yeomans DK (2013) *Near-Earth Objects, Finding Them Before They Find Us*, Princeton University Press, Princeton, NJ.

Index

Boldface indicates main reference

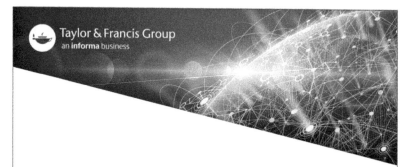